AF247460

The High-Tech War of Twentieth Century

The High-Tech War of Twentieth Century

VISHWA MOHAN TIWARI
AIR VICE MARSHAL (Retd.)
and
Dr RAJNI KANT TEWARI

VIKAS PUBLISHING HOUSE PVT LTD

VIKAS PUBLISHING HOUSE PVT LTD
576 Masjid Road, Jangpura, New Delhi 110 014

First Floor, N.S Bhawan, 4th Cross,
4th Main, Gandhi Nagar, Bangalore - 560 009.

Branches:
NEW DELHI ● BOMBAY ● BANGALORE
MADRAS ● CALCUTTA ● PATNA ● KANPUR

First published, 1996

ISBN 81-259-0159-0

Printed at Ram Printograph, Delhi - 51

Dedicated to dear friend
Air Marshal Vir Narayan PVSM

Contents

War is a Serious Business, During Peace Time Also
Man is a Technological Animal
Role of Technology in War
Cannons Raze Forts to Ground
Birth of Modern Tactics
Rifles Defeat Guns
Impact of Technology on War is Very Complex
Offence is the Best Policy, Not Always !
Technology Needs a Study in Depth
Air Power in WW I
Technology in WW I
To Attack or to Defend, is the Question
Dare Devil Pilots
Role of Tank in WW II
Leader Falls Behind
Electronic Warfare : War With Invisible Rays
Radar, the Penetrating Eye
Diffident France Dilli-Dallies
Seed Sprouts in Other Countries

Technology Shall Prevail
Battle of Britain, Battle of Brains
Electronic Warfare is Technology and Brains
Billion Dollar Atomic Bomb

Birth of Kuwait and Iraq
Kurds get a Raw Deal
Kuwait, Oil and Iraq
Military Power Rules. O.K.
Iraq's Economy is Shattered
Guns Yes, Bread No
A Journalist is Hanged
Saddam Threatens
Grand Plans
Incidences Foretell
Saddam's Size
2nd August, 1990
A Dagger Thrust into Arab Brotherhood
Quickest Victory: Longest Misery
Righting the Wrong Begins
Desert Shield is Created
Drive the Aggressor out: UN Mandate
Saddam Ties the Hands of his Army
Attacker is Attacked

Gulf War on TV
Questions TV Did Not Answer
How did the MNF Achieve Air Superiority
Air Power
Air Defence by Iraq
Air Power Concepts of Iraq
Radar, an Electronic Eye
Surface-to-Air Missiles (SAMs)
Electronic Order of Battle
Airborne Warning and Control System (AWACS)

<u>FOREWORD</u>

1.   It gives me great pleasure to write the foreword for this book which reflects the ubiquity of air power in any modern confrontation. While a lot has already been said and written on the Gulf War, it continues to evince keen interest among professional air power analysts.

2.   This is an ideal book for a layman looking for more detailed account of the Iraq-MNF confrontation. The "Hi-Tech War of Twentieth Century" is neatly packaged to keep the interest of the reader alive. The history of Iraq and Kuwait as also the balance of forces provides the necessary backdrop and events leading to the conflict. Unfolding the various aspects of the short war in a very systematic manner, the author keeps the reader engaged in a style which is easy going and not too technical. Further, the role of technology in warfare as also some historical imponderables have been analysed. Air Power enthusiasts need to take note of the importance of uptodate intelligence and target information, electronic warfare and high technology weapons, which when combined and integrated with joint ops plans provide the correct recipe for unqualified success.

3.   All the same, there are several lessons for military analysts and for readers alike, considering that these events give a clear indication of the kind and manner in which future battles are likely to be fought and the role destined for Air Power.

AIR CHIEF MARSHAL
CHIEF OF THE AIR STAFF

29 Aug 95

अथ चेत्त्वमिमं धर्म्यं संग्रामं न करिष्यसि।
ततः स्वधर्मं कीर्तिं च हित्वा पापमवाप्स्यसि॥

गीता २/३३

If thou doth not fight this moral war, thou shall
lose, not only fame but also thine raison dètre
and be a sinner.

Geeta 2/33

Preface

It must be admitted that 'to war is human, and to avert it is divine'. If physically weak homo sapiens are ruling the earth, it is partly because of their ability to develop and use weapons and partly because of their pugnacious nature. Clausewitz, a prominent military thinker, had warned about 200 years ago, "The fact that slaughter (as in war) is a horrifying spectacle, must make us take war more seriously and not provide an excuse for gradually blunting our swords in the name of humanity". Until there is an effective world government, there would be wars between nations. To avoid a war, one must be fully prepared for a war so that any aggressor gets convinced that his efforts would be too expensive in terms of men and material.

In a democracy , many a time, preparedness for a war does not get sufficient priority. Spending money for preparation of a war, which cannot be seen on the horizon, is not so effective for catching votes as for any other development work. Therefore unless the people themselves realise the value of defence preparedness, the politicians would not give it due importance till the war clouds can be seen darkening the horizon. But in today's hi-tech military environment it would be too late, because the aggressor would be well prepared and one cannot prepare oneself for a hi-tech war in a short period. Preparedness for a hi-tech war requires continuous hard work, specially so for not so advanced nations. Hence there is the need for 'popular' books on war which can create a balanced and sane environment for defence preparedness.

Many books have been written on many wars including Gulf War - 91, but none of them has satisfactorily highlighted the crucial importance of high-tech weapons, their impact on strategy as well as tactics etc. If the general public gains some knowledge on matters of 'Defence' and the importance of hi-tech in it, not only more able persons would like to join the Forces to try their capability and creativity in solving problems but also the defence budget would not remain a sacred but dumb cow, and 'Defence' may get serious consideration that it deserves. An important reason for the lack of popularity of military history is the belief (myth ?) that war is an aberration in the historical process and that, consequently, the study of war is inconsequential. Indeed, the more one studies war, the more one gets a deeper insight into human nature.

In democracy it is essential that public opinion is built on knowledge and wisdom. It was therefore considered necessary to write a book on a modern hi-tech war to enlighten the public. In today's hi-tech environment, development of weapon systems, electronic counter measures and counter-counter measures are taking place at an accelerating pace. Therefore regular and proper funding for war preparations is essential.

History can seldom give us total answers to present problems, at least we do not seem to learn correct lessons. The past stubbornly refuses to yield direct lessons for the present. Then why read history, at all, of any war ?. Edward Mead Earle had stated that, " a knowledge of the best military thought will enable readers to comprehend the causes of war and its fundamental principles". He added that," eternal vigilance in such matters is the price of liberty and for a durable peace a clear understanding of war is essential".

Gulf War-91 was seen by people in their drawing rooms, day by day, courtesy Peter Arnet and his channel CNN. Naturally the details of the war were discussed over cups of tea. This show on TV must have kindled an unprecedented curiosity in the common man and many questions that must have cropped up in his mind could not have been answered by the electronic media due to its obvious limitations.

This war should have been aptly called 'Kuwait War'. However, with Gulf oil predominant in the minds of Westerners, the newspapers and electronic media called it 'The Gulf War'. Therefore, we are also using the term 'Gulf War' but the term 'Kuwait War' has also been used in the book whenever called for.

We have also attempted to answer some of the questions that a keen mind would have asked when the war was being fought. We faced many difficulties in gathering necessary information, specially because nothing was available from any independent authentic source on the strategy of Iraq. This is not to say that all the information was available on the strategy followed by the Multi-National Forces (MNF). The problem here was different, it was to sift information from disinformation. The data on the strengths of various armies and air forces may also be taken as reasonable estimates, although different sources have given different statistics. We believe that this approximation would not dilute the conclusions drawn or weaken the analysis.

Gulf War-91 is the most recent war fought with latest hi-tech weapons. It has been shown in the book that Gulf War 91 is the most hi-tech war of twentieth century, not only because latest hi-tech weapons were used in this war, but also because these weapons were used intelligently and decisively. It is not that in WW II or WW I hi-tech weapons of their age were not used, but that their real impact and purport was not well understood and at times, even misunderstood.

Technology today affects, more deeply than ever before, almost all aspects of war e.g. military organisation, strategy, tactics, concentration of forces, flexibility, logistics, preference between offence and defence, choice of weapon systems and targets surprise, security and morale. After studying various wars, it appears that role of new weapons, product of hi-tech of the time, has not been easy to understand. Of course, as always, there were military thinkers for and against a new weapon. Probably the subject of warfare has become even more technical and therefore Generals themselves either don't have time for it or don't want to get into the complexity and depend upon advice of some experts, and then ultimately play safe.

Apart from many lessons that come out of " The Hi-Tech War of Twentieth Century", the necessity to appreciate the impact of technology on war comes out loud and clear. Technology, apart from being an expert's domain and being complex, is also very dynamic in nature. The first tank (tracked armoured gun) was demonstrated to military Generals, as early as in 1903 in France, 1908 in Britain and 1911 in Germany, but the Generals rejected it because the speed and range of the tanks, at that time, were very low. The Generals could not envisage that with more research and development work, the tank could become an effective weapon. It was during early 1915 in the thick of WW I, when trenches in combination with machine guns brought the two armies to a standstill, that the vital need of a tank was realised. Generals then wanted them post-haste. Fortunately, the Allies produced tanks earlier, trampled the 'trench cum machine gun' defence and marched to victory. Such cases have kept recurring till date. Despite the phenomenal success of electronic warfare in WW-II, many developing countries have not realised the true importance of electronic warfare. Sometimes it had been because of lack of expertise in the field, sometimes because of rivalry between soldiers and technologists. Senior soldiers often find the subject difficult to grasp because of its abstract and complex nature. It is therefore essential to have a team of technology experts, military thinkers and 'soldiers' to discuss the role of technology and its probable impact on war for the next five to ten or even twenty years.

Democracy needs commitment, knowledge and wisdom in many fields, otherwise it can be hijacked by demagogues and criminals. One of the main aims of this book is to produce a sane awareness about war.

The subject of war has become doubly technical - firstly, military science and art; and secondly, technology & war. Generally, in books on war for an informed common man, technicalities are avoided and concepts etc. are just mentioned. We have decided otherwise and given brief technical details. If a reader understands them, he is not required to look for a specialised book to appreciate the book in toto. The subject of electronic warfare is difficult, complex and abstract, but it is not new. The subject of information management, specially for war, is in its infancy but is growing at a precocious rate. Its impact is also not easy to grasp. We have attempted to give some detailed knowledge on the subject. Most of the readers may like just to glance at it to get some idea about its complexity. If a reader does not understand an aspect, despite the help of technical glossary provided, he should carry on reading and yet he would get the 'main story' of the book. If a reader has got a background of science education, he will find the subject somewhat easier to grasp. A non-science background reader would also get benefitted if he accepts various technical concepts and descriptions while reading the book.

It is hoped that serious students of war would find the book interesting, informative and thought provoking. We further wish to inform the readers that Chapter-I and indeed the whole book should be read as one feels inclined towards it, e.g. readers not interested in the technicalities of the chapter may please read it at a fast pace and then may return to it after reading the entire book, if felt necessary. A serious book deserves a second reading, if full value is to be obtained.

If readers were to communicate their reaction to the book, the authors would feel happy and get encouragement for writing more books on the subject of Defence', specially so, during these days when the written word is under attack by the 'television'.

Vishwa Mohan Tiwari
Rajni Kant Tewari

"To war is human, and to avert it is divine"- V.M.T

वाय उक्थेभिर्जरंन्ते त्वामच्छा जरितार: । सुतसोमा अहर्विद:।

ऋग्वेद 1/2/2

Wind has the power to move. Wind is the efficient cause of technology. Savants who produce brain tonics from herbs, and obtain knowledge of science, and whose knowledge is respected for its scholarship, applications and beneficial effects for all, throw light on virtues of wind and sing hymns in the praise of wind.

अश्विना यज्वरीरिषो द्रवत्पाणी शुभस्पती । पुरुभुजा चनस्यतम् ।

ऋग्वेद 1/3/1

Water and Fire, the two Ashwins, with their high speed and energy, transform knowledge in to practical applications; these Ashwins are the patrons who throw light on the processes of building and construction; also these Ashwins help in production of different items for consumption. Let learned persons invoke the help of Ashwins to perfect their art and science of construction, to produce machines and aircrafts etc; and through these obtain happiness.

The mantras have been taken from Rrigveda Bhashya Bhaskar. Bhashya is by Sayan, introduction by Swami Dayanand and interpretation by Pt. Sudarshan Dev Acharya (Book Published by Arsh Sahitya Prachar Trust, 455 Khari Baoli, Delhi - 110006)

Acknowledgements

We have obtained most of the information from various issues of Aviation Week and Space Technology, Jane's Defence Weekly, Air Clues, Vayu, Journals of United Services Institute of India (USII), IDSA (Institute for Defence Studies and Analysis), Sunday Telegraph Magazine, Time Magazine, Wireless files from USIS, Military Balance, Business Standard (India) etc. Historical material has also been collected from various volumes of Encyclopedia (15th Edition); Makers of Modern Strategy (Editor- Peter Paret), Electroniki Yudh Kala (Author-Vishwa Mohan Tiwari), A Study of War (Author- Wright), Khari Yudhha 91-Vayu Shakti Ki Vijay (Author-Vishwa Mohan Tiwari), 'Decision Support Systems in Defence' (Thesis by Wg. Cdr. Rampal), 'An Information War' (A paper by Monica Chikermane) and many others. We wish to convey our deep gratitude to all these magazines and books. We are also thankful to the USII library where we got most of the magazines, and to the USIS for providing us some very useful photographs along with the related information.

The authors are extremely grateful to Air Chief Marshal S.K. Kaul, PVSM, MVC, ADC, Chief of the Air Staff for having very kindly agreed to read through the manuscript, writing the "Foreword" and for offering many pertinent suggestions. We profusely thank Dr. Shyama Charan Dubey for his invaluable guidance and encouragements. The valuable comments by Air Marshal Vir Narain PVSM are also gratefully acknowledged. Thanks are also due to Sh. Som Nath Chaudhary for his help during preparation of the manuscript.

Finding appropriate Vҽdic mantras, befitting the title of the book, was proving to be an extremely difficult task till Shri Ambarish Buddhiraja, a Vedic scholar was approached. We are grateful to him for suggesting quite a few 'mantras' with their authentic interprtations.

Important Historical Events of Iraq and Mesopotamia

BC

10000 - 3000	Neolithic and Chalcolithic Age.
2700 - 2350	Sumerian Kingdom.
2350 - 1794	Akkadian and Isin Dynasties.
1794 - 1600	Babylonian Empire, Hammurabi, Samsuiluna, Samsuditana, Kassaites, Hittites etc.
1600 - 1450	Dark Period, Hittites, Kassaites, Hurrians, Mittani etc. (probably, all Aryans)
1450 - 1155	Kassaites, Hurrians and Mitanni Kingdoms.
1156 - 1025	Second Isin Royalty.
1024 - 1004	Second Kassaite Royalty.
1003 - 627	Neo-Assyrian Empire.
626 - 539	Neo-Babylonian Empire.
539 - 335	Iranian Empire.
335 - 142	Alexander, Selucids, Parthian Empire.
141(BC)- 113(AD)	Parthian Empire.

AD

114 - 117	Roman Empire.
118 - 636	Roman, Parthian, Sasanian Kingdoms.
637	First Arab King, Saad Ibn Abi Vakkas. Iraq becomes a geographical expression for the flat lands between Baghdad and the Persian Gulf. Iraq means a country with healthy roots.

658	Crushing of Kharijites by Ali.
661	Ali killed by a Kharijite, Syrian Governor Mu'awiyah shifted the seat of power of Khalifa to Damishk (Damascus), Iraq was made a province of Syria.
680	Mu'awiyah died. Al Hussain (second son of Ali) rose against Kharijites' kingdom but was killed.
749	Mutiny by Abbasids against Umayyads and Kharijites, and the first Abbasid Khalifa installed as the King.
749 - 933	Rule by Abbasid Khalifas.
934 - 1055	Buyids Kingdom.
1055 - 1194	Rule of Seljuq Turks.
1194 - 1258	Opposition by An-Nasir and sons.
1258 - 1410	Mongol Rule.
1411 - 1508	Turkmeni Rule.
1508 - 1534	Safavid (Iran) rule.
1534 - 1917	Ottoman Empire.
1918 - 1921	British Rule.
1920	Mosul, Baghdad and Basra merged by the British to establish the present Iraq.
1922	Kurds established an independent state in Iraq but were crushed by the British.
1921 - 1932	Keeping the foreign policy with themselves, Britain declared Iraq an independent country. Declaration of Faisal I as the first King of Iraq Republic.
1923	Iraq-Kuwait border agreement under the auspices of British.
1932	Complete freedom to Iraq.
1932	Abdel-Aziz Al Saud unified four Bedouin tribes and created Saudi Arabia.
1932 - 1935	Formation and fall of eight Governments in quick succession. The 1923 Boundary Agreement confirmed through the Memorandum of Iraq-Kuwait Convention on Boundaries.

1936	With the help of armed forces, Hikmat Suleman staged a coup against the Government and became its Prime Minister.
1939	King Gazi died after 6 years. Gazi's son Faisal II crowned. Brother Amir Abd-al-Ilah became the Regent. General Nuri became the Prime Minister.
1940	Rashid Ali became the Prime Minister and was unwilling to cooperate with Britain and began secret negotiations with Axis powers.
1941	British attacked Iraq successfully during early 1941; Rashid Ali fled. British dominated Iraq during WW II.
January 1946	Formation of a new Government.
May 1946	General Nuri became the Prime Minister.
March 1947	Salih Jabr became the Prime Minister.
1948	General Nuri again became the Prime Minister.
1951	Actual demarcation of boundary was requested by Kuwait. Iraq asked for Burbah island as a bargain.
1953	After attaining adulthood, Faisal II became the King.
July 1958	Brigadier Abd Al Karim Qasim occupied Baghdad with the help of army, Iraq was declared a democracy, King assassinated, Council took over the power.
7 October 1959	Plot to kill Qasim by six member squad, including Saddam Hussein, failed. Saddam fled to Egypt.
25 June 1961	Qasim declared Iraqi occupation of Kuwait, title of 'Qaim Maqam' given to Kuwait Sheikh.
February 1963	Mutiny by Ba'th Party with the help of army, assassination of Qasim, Abd as-Salam Arif became the President, Revolutionary Command National Council formed, and

	took over the legal and administrative authority. An officer of Ba'th Party, Col. Ahmad Hasan Al Bakr became the Prime Minister.
	Saddam was made a member of the Regional Command Council and was incharge of internal security.
4 October 1963	Iraq recognised Kuwait. In a separate agreement, Iraq agreed to provide Kuwait with 120 million gallons of water per day from Shatt al-Arab in exchange for lease of Bubian island.
November 1963	President Arif arrested the members of Ba'th Party with the help of Armed Forces and took over complete authority.
1964	Declaring the 1963 Agreement as invalid, Iraq claimed some parts of Kuwait during the Boundary Committee Meeting.
September 1965	Arif made Abd ar-Rahman al-Bazzaz the Prime Minister.
April 1966	President Arif died in a helicopter accident. His brother Abd ar-Rahman Arif became the President. Abd ar-Rahman al-Bazzaz resigned and an officer from the army became the Prime Minister.
July 1968	Mutiny by Ba'th Party with the co-operation of armed forces. Arif killed. A member of Ba'th Party, Ahmad Hasan al-Bakr became the President. Chairman of the R.C.C., Secretary General of the Ba'th party and C in C. Saddam was made the Deputy Chairman of the R.C.C. and in charge of internal security.
1969	Shah of Iran refused to pay tax for Iranian vessels passing through Shatt-al Arab. Iran and Iraq started propaganda war. Iran started assisting Kurdish independence movement.

1971-72	Iraq asked for Bubian island in exchange for supplying water to Kuwait.
20 March 1973	Iraqi army occupied a Kuwaiti post Al-Samitah but later vacated under pressure from Saudi Arabia and Iran.
August 1973	Iraq offered to accept the boundary if Barbah & Bubian islands were given on lease or otherwise.
1975	Propaganda was stopped, Iran stopped aiding Iraqi Kurds; and Iraq agreed to allow the passage to Iranian vessel through Shatt al-Arab.
July 1979	Saddam Hussain became the President.
1979	Fall of Shah of Iran.
1980	Gulf Cooperation Council formed by Saudi Arabia, Kuwait, UAE, Oman, Bahrain and Qatar
Feb 1980	Saddam issued a Pan Arab Charter calling for collective Arab defence and prohibiting any Arab state resorting to armed force against any other Arab state.
1980 - 1988	Iran-Iraq war.
22 Sept 1980	Iraq attacks Iran without any warning.
30 Sept 1980	Iran attacks the Iraqi nuclear plant at the Tammuz 17 complex but the damage is marginal.
Sept 1980	Security Council called for a ceasefire through Resolution 479.
7 June 1981	Israeli fighter bombers destroy the Iraqi nuclear plant at the Tammuz 17 complex.
3 June 1982	Israeli ambassador to Britain was shot dead by an Iraqi Secret Police officer. Some analysts believe this was to induce an Arab-Israeli war with an aim to generate a feeling of unity among Arabs and Persians.
10 June 1982	Saddam declared a unilateral cease-fire. Iran did not accept it and continued the war.

1984	Environemnt as a weapon was used by Saddam against Iran when he put on fire the oil wells of Abadan (probably for the first time in history).
1984	Elections for Iraqi National Assembly conducted (single party).
17 May 1987	An Iraqi Mirage F-1 fighter bomber fired two Exocet anti-ship missiles at a US frigate off Iranian coast. Some analysts believe, this "mistake" was enacted to make the presence of US military more effective in the area. As expected, more authority was given to the commanders of the US ships in the area which resulted in a heavy loss to Iranian navy.
18 July 1988	Ayatullah Khomeini accepted the UN sponsored cease-fire.
August 1988	Iraq and Kuwait agreed for the formation of a joint committee for their boundary demarcation. Again Iraq asked for Bubian Island and Kuwait refused again.
October 1988	Iraq forces attacked and entered 20 Km inside Kuwait and withdrew under pressure from world opinion.
November 1988	Saddam promised to bring democracy.
February 1989	Arab Cooperation Council formed with the initiative of Saddam along with Egypt, North Yemen and Jordan.
April 1989	Elections for Iraqi National Assembly took place.
September 1989	Glassnost overtakes Eastern Europe and USSR.
August 1988 to March 1991	Please see the events under 'Kuwait' and 'Kuwait War'.

6 March 1991	Saddam appointed Ali Hassan Majid as Interior Minister. Majid was known as the "butcher of Halabja" because in 1988 he had used chemical weapons against Kurd rebels in which more than 5000 innocent Kurd men, women and children died a most inhuman death.
9 March 1991	Battle between Shi'ite rebels and Republican Guards became serious in Southern Iraq.
20 March 1991	Two Iraqi Su-22 fighter bomber aircrafts took off , probably to attack Kurd rebels, in violation of the cease fire terms. Two E-15A US Air Force fighters engaged them and shot one of them, the other landed back immediately.
23 June 1991	A UN team was denied, after earlier agreement, inspection of Abu Gharaib nuclear installation. Probably, after removing sensitive apparatus, they allowed the inspection on 26 June 1991 where nothing was found. The satellite pictures showed movement of certain items, proving Iraq's intention of not allowing an effective inspection.

Important Historical Events of Kuwait

1700	Aniza tribe from Arabia settles in 'Kut', later known as Kuwait.
1756	Sheikh dynasty established by Sheikh Bin Jabir in Kuwait.
1776	A post office started by East India Company.
1792	The Company shifted its agency from Basra to Kuwait.
1821	British Political Officer installed in Kuwait.
1821	Kuwait was made subordinate to Basra province under the Ottoman Turk Empire.
1871	The Turkish Governor of Baghdad, Midhat Pasha gave the Sheikh of Kuwait the title of Qaim e Maqam (Dy. Governor).
1896	Sheikh Mubarak revolted against Ottoman empire and killed his half brother Sheikh Mohammed who was pro-Turkey, declared Kuwait independent and sought protection by Britain.
1899	Britain's conditional agreement to protect Kuwait and taking - over of Kuwait's foriegn affairs.
1913	Britain got monopolistic rights for oil exploration in Kuwait.

1914	Britain established a protectorate in Kuwait as an independent state.
1921	Sheikh raised the question of demarcation of boundaries with Nezd (Saudi Arabia).
1922	Nezd-Kuwait and Saudi-Iraq boundaries agreed - Al Dakreiyar Agreement.
1923	Kuwait-Iraq boundary agreement with the intervention of Britain
1932	Through the Memorandum of Iraqi-Kuwait Convention on Boundaries, the 1923 Boundary Agreement confirmed. Iraq sends a memorandum to British Resident located at Kuwait.
1938	Oil exploration in Kuwait successful.
1951	Iraq asked for Burbah Island in return for Kuwait's request for boundary demarcation with Iraq.
19 June 1961	Britain cancelled all previous agreements and declared full freedom for Kuwait.
25 June 1961	Iraqi Prime Minister Qasim declared Kuwait as a part of Iraq. The reason given was that like Mosul anu Baghdad, Kuwait was also a part of Basra.
20 July 1961	Kuwait being independent country, Arab League gave Kuwait its membership.
14 May 1963	Kuwait became a member of United Nations.
04 Oct 1963	Iraq Government recognised Kuwait as an independent country.
1964	Declaring the earlier agreement as invalid, Iraq claimed some parts of Kuwait during joint Kuwait-Iraq boundary committee meeting.
July 1968	Ba'th party declared its authority over Kuwait after a successful coup in Iraq.
1971-72	In lieu of supply of water to Kuwait, Bubian Island was asked for. Kuwait refused the demand.

20 March 1973	Iraqi Army attacked and captured a Kuwaiti post, Al-Samita, but vacated under pressure from Saudi Arabia and Iran.
28 April 1973	Iraq suggested to Kuwait that apart from other issues they should take the larger interest of Arab world into consideration during further discussions on the boundary problems.
August 1973	Iraq mentioned that they will agree on the present boundary provided Kuwait gives them Burbah and Bubian islands on lease or otherwise.
1977	The bilateral meeting conducted to solve the boundary dispute unsuccessful due to the unjustified demands of Iraq.
1980	Kuwait gave financial, political and material aid to Iraq during the Iraq-Iran war, Sep 1980 - July 1988.
1984	A report came that the two countries have reached an agreement that the islands of Burbah, Bubian and Failaka will be kept under Iraq's control for their safety.
August 1988	Iraq and Kuwait agreed for the formation of a joint commitee for their boundary demarcation. Again Iraq asked for Bubian Island and Kuwait refused.
October 1988	Iraqi Forces attacked and entered 20 km inside Kuwait. Withdrawal after discussions and world opinion.
February 1989	Prince and Prime Minister of Kuwait, Sheikh Saad Abdullah reached Baghdad but could not solve the boundary dispute.
April 1990	A new National Assembly constituted as part of democratisation process.
16 July 1990	Iraqi Foreign Minister wrote to Arab League blaming UAE and Kuwait for occupying Iraqi territory, theft of their oil and

	extraction of more oil than agreed. The last complaint was not called for as Kuwait and UAE had already agreed on 10 July 90, at a ministerial level meeting of OPEC at Jeddah.
24 July 1990	President Hosni Mubarak of Egypt arrived at Baghdad for intervention. He suggested for a meeting of Arab Foreign Ministers in Cairo. Iraq concentrated its forces on the Kuwait border but Saddam assured Mubarak that he would not attack Kuwait.
24 July 1990	A joint exercise at short notice was held between the US and the UAE with warship and aircrafts in the Persian Gulf, 960 Km south- east of Kuwait.
25 July 1990	Iraq declared that it would not be intimidated by the US tactics in the Persian Gulf.
31 July 1990	Iraq and Kuwait consented for peace talks in Jeddah. The talks broke down the very second day due to unjustified demands of Iraq.
2 August 1990	At 0200 hrs Iraq attacked and occupied Kuwait within 4 hours.

Events of the Kuwait War

21 July 1990

About 30,000 Iraqi troops positioned along the Kuwaiti border.

24 July 1990

The US held a joint exercise with the UAE deploying warships to patrol the northern gulf.

26 & 27 July 1990

In the OPEC oil Ministers meet in Geneva Iraq proposed the price of oil to be raised from $18 to $25 a barrel. A compromise price of $21 was agreed by all including a happier Iraq. But the troops continued to mass on the border.

28 July 1990

Bush sent a message to Saddam stating that the use of force to resolve Iraq grievances against Kuwait would be unacceptable and the US would support its friend in the region. This was a result of photographic evidence given by satellites that around one lakh Iraqi soldiers were deployed at the border and CIA had come to the conclusion that Iraq is likely to attack.

31 July 1990

A meeting between Iraq and Kuwait took place at Jeddah. Iraq declared that the Kuwaities did not want a settlement, whereas Kuwait stated that it had agreed to write-off the Iraqi debts and lease Warbah Island to Iraq and was ready for further negotiations.

31 July 1990

Saddam invited western military attaches to visit the border and find out for themselves the truth about the imminent attack . They reported that the Iraqis were not equipped for action.

1 August 1990

In a meeting in the White House, National Security Agency of the USA indicated that it was sceptical of the CIA's alarmist reports, whilst the CIA maintained that the attack will take place.

2 August 1990

Iraqi Forces attacked and captured Kuwait without any significant resistance from the too small Kuwaiti Army. Indeed the capture was so peaceful that the people discovered it only the next morning when they woke up. The Emir of Kuwait fled to Saudi Arabia.

Emergency meeting of UN Security Council held and with 14-0 votes, under Resolution 660, Iraq was considered culprit and was asked to unconditionally vacate Kuwait. President Bush ordered economic sanctions and trade embargo with Iraq, and moratorium on the wealth of Iraq and Kuwait in USA.

3 August 1990

Iraqi Forces march towards Kuwait-Saudi border. Bush gave strong warning to Saddam against his aggression on Saudi Arabia. Saddam's declaration for a meeting with Emir of Kuwait within two days, and promise to start vacating Kuwait immediately after the talks. Belgium, France and Britain also put moratorium on the Kuwaiti wealth in their countries.

4 August 1990

Information regarding concentration of Iraqi forces in Kuwait received through satellite photographs. Canada, Japan and European Community imposed embargo on oil from Iraq and Kuwait.

6 August 1990

Iraqi soldiers arrested American and British citizens and kept them as hostages in Iraq. King Fahd of Saudi Arabia invited friendly nations to defend his country. President George Bush then ordered an F-15 squadron and 82nd Air Transportable Division to reach Saudi Arabia. With 13-0 votes the Security Council ordered for economic sanction and trade embargo against Iraq.

7 August 1990

Saudi Arabia closed the Yanbu pipeline of Iraq. Turkey closed the Mediterranean pipeline of Iraq.

8 August 1990

Iraq declared Kuwait as its nineteenth province. An Arab summit in Cairo, endorsed the Saudi Arabia's decision to call in 'multi-national forces' help to defend Saudi Arabia, among other objectives. Bush declared the operation 'Desert Shield'.

10 August 1990

13 out of 17 countries of Arab League and all the countries of Gulf Co-operation Council declared to send their Peace Forces to Saudi Arabia. Member countries of NATO also declared their help.

11 August 1990

Saddam declared that he would vacate Kuwait (as per UN Resolution 660) if Israel would vacate occupied Arab lands (as per UN Resolution 242).

15 August 1990

Surprisingly, Iraq withdrew his troops from Shatt al-Arab, probably his only booty of the longest war and also as a gesture of friendship to Iran.

16 August 1990

President Bush, after meeting King Hussain of Jordan, stated that Jordan would abide by the restrictions imposed by the UN Security Council.

22 August 1990

For the first time since World War II, Bush activated the military reserved units.

28 August 1990

Iraq gave constitutional validity for Kuwait being its nineteenth province.

13 September 1990

Japan declared US $4 b aid to Multi-National Forces.

23 September 1990

Iraq warned to wage a war against the Multi-National Forces if the UN trade embargo results into their starvation.

3 October 1990

At UNO, the Foreign Ministers of Organisation of Islamic Conference condemned the Iraq's aggression. Amnesty International blamed Iraqi forces for torturing the Kuwaitis in Iraq.

4 October 1990

UN representatives and Foreign Ministers of more than hundred neutral countries demanded Iraq to vacate Kuwait.

20 October 1990

The International Parliament Union requested its members and the Governments to help in executing all the resolutions of Security Council. In its meeting at Montevideo, Uruguay demanded Iraq to withdraw its army from Kuwait.

9 November 1990

The General Committee of the UN General Assembly rejected Iraq's contention that the presence of American forces in the Gulf is a threat to Arab as well as to the international peace. Contrarily the members felt that Iraq is a threat to the area.

24 November 1990

Member nations of SAARC declared that the Iraqi forces should unconditionally withdraw from Kuwait immediately.

26 November 1990

Five permanent members of Security Council gave their consent in principle for using force against Iraq if it does not vacate Kuwait. President Gorbachev, while addressing the Supreme Soviet, warned Iraq that it would be punished for its aggression.

29 November 1990

The Security Council authorised Allied Nations to take suitable action against Iraq to enforce the charter of Resolution 660, if Iraq does not vacate Kuwait by 15 January 1991, and passed the Resolution 678 to that effect.

30 November 1990

Iraq rejected the Security Council Resolution 678 labelling it as illegal and unacceptable.

2 December 1990

Presidents of five countries wrote to Saddam for implementation of Security Council Resolutions.

5 December 1990

Iraq accepted President Bush's suggestion for a meeting at the highest level.

6 December 1990

All the hostages released by Iraq.

14 December 1990

Bush warned Saddam that he should give his consent to meet James Baker, US Secretary of State, before 3 January 1991 in Baghdad.

15 December 1990

Iraq declared that its Foreign Minister, Tariq Aziz would not go to Washington on 17 December, the earlier suggested date. A fresh date would be fixed by Iraq and conveyed to US.

18 December 1990

With a large majority, UN Assembly strongly condemned the Iraq's violation of human rights of citizens of Kuwait and other third world countries.

22 December 1990

Saddam threatened to use chemical weapons if Kuwait or Iraq was attacked.

26 December 1990

Soviet Union sent two high level representatives to Iraq for discussing the vacation of Kuwait.

3 January 1991

Bush said that he is prepared to send his Secretary of State, James Baker, to Geneva for meeting the Foreign Minister, Tariq Aziz of Iraq.

The Foreign Ministers of Pakistan, Iran and Turkey, in a meeting at Islamabad, addressed Iraq to vacate Kuwait.

4 January 1991

Iraq consented to send Tariq Aziz to Geneva. EC foreign ministers accepted, informally, in principal, that an international conference be called to discuss all problems of the Middle East, including the Palestinian issue, in return for Iraqi withdrawal.

6 January 1991

On Iraqi Army Day, Saddam, declared with his usual bravado, that all the issues of Middle East were part of one battle and the brave Iraqi people were ready for "the mother of battles".

8 January 1991

Baker and Aziz reached Geneva for the meeting. Bush requested the American Congress to ratify all the resolutions of Security Council including the Resolution No. 678.

9 January 1991

Discussions between Baker and Aziz took place for six and a half hours. Baker later told that Iraqis did not show any desire to implement the resolutions of the Security Council. On this Bush said he would support a meeting between De Cueller, the Secretary General UNO and Saddam. Secretary General agreed to go to Baghdad for the meeting. Tariq Aziz condemned the double standards of the West, in allowing Israel to ignore the UN resolutions.

12 January 1991

The US Congress authorised President Bush to use American forces to implement the Resolution 678 of Security Council.

13 January 1991

After meeting Saddam for 90 minutes, the Secretary General of UN said that he could not succeed in convincing Iraq. The Secretary General informed Bush that there was no need for another Security Council meeting before initiating military action.

14 January 1991

The Secretary Generals of Organisation of Islamic Conference and African Unity Organisation, acting President of Bangladesh and the Australian Cabinet etc. appealed to the President of Iraq to abide by the Resolution 678.

15 January 1991

The UN Secretary General again sent an appeal to Iraq.

The US Ambassador to UNO said that Iraq, hopefully, would relent to De Cueller, otherwise Multi-National Forces were ready.

The German Parliament took note of the fact that Iraqi leaders have refused to vacate Kuwait and thus closed the doors for peaceful settlement.

The Italian Prime Minister, Julio Andreotti said that, on principle, this is not a matter between Iraq and US but it is between Iraq and UN.

The French Prime Minister, Micheal Rokar blamed Baghdad that Iraq had rejected all the efforts to solve this problem through negotiations and said that now it is time to act.

16 January 1991

The Multi-National Forces (MNF) attacked Iraq on midnight under the operation 'Desert Storm

25 January 1991

Iraq started pumping oil from the wells into the Gulf.

28 January 1991

World Health Organisation (WHO) told Iraq that pumping oil into the Gulf is a grave threat to environment.

The MNF attacked the Kuwaiti oil wells which stopped the oil being pumped into the Gulf.

31 January 1991

Iraqi tanks took an offensive and rolled into Khafji. It took considerable air support and 30 hours when the MNF could drive them back. The loss of MNF soldiers was one of the heaviest of the Kuwait war viz. 50 Saudi soldiers were killed or badly wounded and 11 US Marines were killed.

15 February 1991

The Government of Iraq declared on Baghdad radio that as per UN Resolution No. 660 they are prepared to withdraw from Kuwait provided MNF also withdraws from the Gulf region and simultaneously Israel also leaves the area illegally occupied by them.

Bush termed the Iraq's proposal a mere bluff and said that Iraq must vacate Kuwait unconditionally otherwise MNF would continue with their attacks.

18 February 1991

Soviets submitted a peace plan :
  i) Iraq should unconditionally vacate Kuwait.
  ii) Moscow would oppose the embargo against Iraq as well as any move to punish Saddam or penalise Iraq.
  iii) The detailed problems of the Middle East should be thoroughly discussed.

19 February 1991

Bush said that the Soviet proposal does not meet the demands contained in the Resolution of Security Council.

21-22 february 1991

Iraq replied positively to Gorbachev's peace plan with,inter alia, one proviso that UN sanctions against Iraq should be lifted after two-third of Iraqi forces had returned.

22 February 1991

Bush declared the soviet peace proposal as conditional and therefore not meeting the Resolution 660 and told Saddam that if he does not want ground battle, he should start withdrawing from Kuwait from 23 February afternoon.

Tariq Aziz discussed the peace plans with Soviets at Moscow.

Iraqi Revolutionary Command Council denounced the President's shameful ultimatum.

Saddam Hussain started putting a series of Kuwait oil wells on fire.

23 February 1991

MNF started ground attack : 'Operation Sabre'.

25 February 1991

Saddam ordered his forces to make a "fighting withdrawal" from occupied Kuwait. USA stated this as insufficient to meet the demands of Resolution 660 and MNF continued their attack.

26 February 1991

Iraq said that they are withdrawing their forces as per the Resolution 660.

27 February 1991

Bush declared on the National TV "Kuwait has been freed, Iraqi forces have lost, MNF should stop attack operations."

2 March 1991

Security Council, with a large majority, passed a resolution to formally end the war. The Resolution contained :

   i)   Iraq should cancel the annexation of Kuwait.
   ii) Iraq should immediately return all the prisoners of war.
   iii) Iraq should return all the property of Kuwaitis and accept the moral responsibility for the loss of property and lives on account of aggression on Kuwait.

6 March 1991

President Bush addressed a joint session of the US Congress, "Today it is dawn of peace in the world". Then he presented the following plan for building a permanent peace in the Gulf region:

   i)   Help from friendly countries for safety of the Gulf region.
   ii) Control of mass destruction weapons and their launching missiles.
   iii) Special political efforts to tackle the differences between Israel and Arabs as well as between Israel and Palestine.
   iv) To evolve a solution for the internal conflict of Lebanon.

14 March 1991

The Emir of Kuwait, Sheikh Zabir Al-Ahmad Al-Sabah returned to Kuwait after seven and a half months as the king of a Sovereign nation.

16 March 1991

Saddam again promised to bring in multi-party democracy.

"Victory in a war depends 99% on hi-technology and 1% on other factors." —— General John Fuller (Britain)

Chapter 1

Technology and War

War is a Serious Business, During Peace Time Also

If 'homo sapiens' are ruling this earth, despite being physically weaker than many other animals and even some hominids, it is because of their ability to use weapons and instruments, and also because of their aggressive nature. The homo sapiens have either destroyed or absorbed the more intelligent and equally robust Cro-Magnon race (brain volume of 1600 cc compared to our's of 1300 cc) most probably because of pugnacious nature of homo sapiens. War is deeply rooted in man's nature. Further as Clausewitz had warned , "the fact that slaughter (as in war) is a horrifying spectacle, must make us take war more seriously and not provide an excuse for gradually blunting our swords in the name of humanity".

Man is a Technological Animal

Technology, definitely, distinguishes man from other animals. The most primitive of men knew the use of a properly shaped 'stick and stone' for both defence and offence. In that he intuitively realized that a sharp edged stone or a stick is a better weapon, he is a born engineer. With competition among 'homo sapiens', engineering also started evolving. Ability to produce fire, to domesticate horse and invent spoked wheel chariot must have given tremendous advantages (as demonstrated by their spread over Eurasia) to the Aryans who harnessed them first. Evolution of war is mainly evolution of weapons and their deployment, strategy, tactics and logistics etc. Just as violence or war is deeply rooted in us, so are love, peace and morality. Indeed we are, most of the time, torn by the opposite pulls of these two deep forces within us. Yet if we wish to avoid wars whilst conducting negotiations we must be fully prepared for it. The strength of negotiations, apart from their justification and forceful presentation also comes, unfortunately, from barrels of guns. Today, we can neither be well prepared for wars nor can we avoid them unless we are capable of employing high technology. Sweden and Switzerland are living examples of this. It is not only evolution of war, but also evolution of culture, agriculture and machines which are intimately dependent on evolution of technology.

Role of Technology in War

The role of technology in war can be divided into five main tasks : (1) Research, design, production, development and maintenance of weapons, ammunitions and fire guidance and control devices (2) Tactical engineering support in a battlefield, e.g. laying or defusing mines and laying communication lines, digging of trenches etc. ; (3) Strategic engineering support, e.g. construction of forts, air fields, arctic tents and fuel dumps etc; (4) Housing and transportation of men and material required for war; (5) Ancillary support such as the preparation, provision and distribution of maps, disposal of unexploded bombs, computerisation etc.

Developments of a stick into a club, into a lance, into a sword are some of the examples of the earliest innovative military engineers most of whom were soldiers themselves. Remnants of hill-forts of iron age, in various parts of the world, are further evidences of constructions carried out by military engineers. 'Vajra' was a high-tech missile weapon of Indra, the king of gods, which could damage forts, considered invincible till then. Invention of an accurate catapult, inter alia, enabled David to defeat Goliath, a good ancient record of victory of high-tech. 'Agni Ban' (fire arrow) was another high- tech weapon of the early Aryans.

Roman engineers also developed, inter alia, a battering ram for breaking the gates of a fort; also a giant catapult and a large crossbow whose bowstring was drawn back by a windlass, used for hurling heavy stones and firebrands etc.

Cannons Raze Forts to Ground

Fire arms and cannons were developed in Europe during the Renaissance. The supreme fine artist Leonardo da Vinci is also an example of his times, par excellence, of a military engineer who refined mundane implements like pick and shovel, and basket and barrow. He developed an engineering skill for casting cannons. He also designed cranes, earth-moving equipment, bridges and assault boats. The art and science of war revolved around besieging and defending fortresses (main functions of military engineers) till 19th century when breech-loading artillery and the high explosive shells were developed which drastically reduced the importance of forts.

Birth of Modern Tactics

Swedish King Gustavus Adolphus had overrun Germany in his reign of two years only (1631 and 32) and is called the father of modern tactics. How did he achieve such a superlative performance? Swedish armour engineers had developed a lighter musket, and to reduce the re-loading time Gustavus introduced use of paper cartridges. Advances in Swedish metallurgy helped in casting four-pounder guns of 3 feet 10 inches in length, of 2.6 inches calibre and only 400 pounds in weight compared to a weight of 1100 pounds or more of other cannons. A single horse or 4 men were sufficient to deploy such a Swedish cannon in the field, a tremendous advantage in mobility and effectiveness of manpower. These technological advantages enabled the innovative Gustavus to reorganise his army for speed, offence and flexibility. He gave two four-pounders to each regiment of 1000 men, for the first time in history.

Marquis de Louvois, Minister of War at the court of Louis XIV of France, foresaw the advantages of two recent improvements in weapons and changed his tactics accordingly. He is therefore known as the founding father of new military age. The first improvement was the socket bayonet which superseded the pike by providing every musketeer with his own means of protection, thereby eliminating the necessity of pikeman for defending the musketeers. The second was the invention of the flintlock musket, making it possible to dispense with clumsy lengths of smoldering "match". Sibastien Le Prestre de Vauban, working for Louis XIV as the Military Chief Engineer, was one of the dedicated supporter of the flintlock musket for the infantry. He was also the inventor of the bayonet with a sleeve or socket that held the blade at the side of the barrel, permitting the rifle to be fired with bayonet fixed. He wrote to Louis XIV recommending strongly the use of flintlocks and bayonets and disuse of pikes. By dedicated work on forts and siegecraft, he became a national idol. He was an honoured public servant, an organisational genius and enlightened reformer. His book, "Seigecraft and the Defence of Fortresses" remained a reference book for many years. He symbolised the new importance of scientific knowledge not only for war but also for the welfare of the state.

Rifles Defeat Guns

Duke of Wellington took advantage of Napoleon's fondness of guns and dislike of rifles. Flintlock rifles, then a recent invention, were superior to muskets in firing-range and accuracy. Wellington very boldly and intelligently positioned his rifleman along with gunners in the front, whereas Napoleonic tactics, till then, was to place guns in front of the infantry along with cavalry and then infantry close behind, and thus Napoleon could control the battle effectively. Napoleon used to blast his way through the enemy coulmns. But with this new rifle, it proved to be suicidal as the gunners were the first targets of the British riflemen; and thus Napoleon met his "Waterloo", to put it rather simply. Even during US Civil War, Gen. Lee at Gettysburg and Gen. Grant at Cold Harbor - the best tacticians of their time - met bloody repulses when they under-estimated the firepower of opposing riflemen. If a General cannot appreciate the effectiveness of a new weapon and can not adapt his strategy and tactics accordingly, he invariably has to meet his 'Waterloo'.

Impact of Technology on War is Very Complex

Offence is one of the ten principles of war and its value in any war is unquestionable. But, it also boosts the ego of a nation or an army or a man and therefore one must be careful, specially when a nation's future or of its fighting soldiers is at stake. Relative importance of offence and defence depends mainly upon geography and weather of the battle area, relative strengths of the military vis-a-vis that of its enemy, urgency, morale, logistics and, last but the most important, technology. The impact of technology, specially high-tech, meaning the latest technology, on the relative dominance of offence or defence, is more difficult to estimate than one may feel intuitively. There are umpteen examples to substantiate this from the invention of rifle, machine gun, tank, aircraft,electronic warfare etc.

'Forts' favoured defence over offence so much that for centuries they extracted heavy price from the attackers. But when powerful cannons came, the advantage that forts offered vanished with a bang, and offence, instead of defence, became the more useful principle of war. Napoleon's successes, to a large extent, depended on his innovative, aggressive and efficient deployment of cannons. He used cannons as the main attack weapon and kept them in the lead to draw maximum benefit out of their firepower. Then came rifles, which had range and accuracy that could attack the gunners. It should have been easy, so one may feel, for any General, let alone Napoleon, to foresee the danger. But rifles produced till then were not a very reliable weapon and therefore their effectiveness was questionable. Duke of Wellington saw the potential and took a calculated risk in putting the riflemen along with gunners in the front row, both in defensive postures as favoured also by the geography of the battle-area. Napoleon's forces, by virtue of being army of the great and powerful nation, had to take offensive measures, whether the situation so favoured or not! Those days the rifle could be used more effectively in defensive rather than offensive deployment as only then it provided the desired accuracy and rate of fire, and therefore could force an effective kill rate on the advancing enemy. Napoleon used to deploy mass cannons as close as 500 yards from the enemy and blast a bloody path to victory. Riflemen of Wellington did not permit the cannons to be brought as close, specially as a low ridge could be used to position the riflemen somewhat safely. Napoleon refused to adapt the rifle into his tactics and neglected musketry training despite his own assertion that "the power of an infantry lies in its fire".

One may think that with such experiences, one may be able to learn lessons and therefore be able to determine whether a new weapon favours 'defence' or 'offence'. The fact that it is not so, demonstrates severe difficulties involved rather than the inability of the Generals.

Machine guns were used as a military weapon by French in the Franco-Prussion war of 1870-71, but they used 25 barrelled Mitrailleuse machine gun like an artillery weapon and therefore failed against breech loading cannons of Prussia. Maxim machine guns first produced in Britain (by an American) in 1883 were also the first successful light weight (70 kg) automatic machines capable of firing several hundred bullets per minute. The Russians used these water-cooled machine guns in the Russo- Japanese war (1904-05), and the Japanese used the French air- cooled machine guns, Hotchkiss. As the effectiveness of a combat infantry depends on the firepower of their weapons- rifle or machine gun- one would conclude that such machine guns would increase the combat potential of a foot soldier dramatically and hence would be welcome by any army. But just prior to WW I, armies could not make up their mind as to how to use these weapons. British rifle- battalions were not given any machine guns but a separate Machine Gun Corps was formed and despite their successes as infantry weapons in WW I, British, surprisingly, continued to give machine guns only to the Gun

Corps. In the German infantry, some of the Companies were equipped with machine guns. The French gave two machine guns to a battalion. This was despite their proven lethality in the US Civil War and in the Russo-Japanese war. What may appear even more surprising is that all the European powers considered machine guns to be a weapon which favoured offence rather than defence. This was again despite the relatively heavy weights of those early machine guns, although the Germans had produced a model which weighed only 45 kg, the lightest in the world during that period. It would be interesting to analyse the psychology behind such a conclusion which today appears to be against common sense. If one has been brain washed by the principle, " offence is the best form of defence", he is looking for an army and weapons which are fit for offence. If he had given machine guns in good numbers to infantry battalions, they would have become less mobile and therefore less offensive. Therefore he did not give machine guns to infantry battalions but either made a separate corps or gave them in very small numbers. Thus he maintained 'offensiveness' in his army but, because of this idée fixe, he could not look at the tremendous potential for defence in a machine gun and could not draw the invaluable advantage that could have accrued nor could he avoid shocking losses of soldiers when attacking a position defended by machine guns.

Offence is the Best Policy, Not Always !

During the early part of the century, both French and German Generals had a faith, which today looks to be more mystic than well founded, on headlong offensive compounded of mass, velocity and morale. Both nations planned for such a speedy offensive action so that the war would not last more than some weeks. No doubt, availability of railways, of light and accurate automatic rifles and armoured cars helped in increasing the speed of advancement or redeployment. The fact that artillery was still mainly horse drawn, did not sober down either Schlieffen or Moltke in their ambitious great wheeling movement to attack France from north-east, after trampling the neutral Belgium, to avoid the strong fortresses built by France on its eastern borders. Although it must be admitted that despite wrong emphasis on offence, the Germans had worked out their offence with more effective tactical planning than the other European countries. The Germans called their new offensive tactics as "infiltration" which meant a rapid advance of dispersed "storm groups" armed with semi-automatic rifles and light mortars, preceding the main body of infantry. At any earlier times, such a tactics would have been considered suicidal but with the production of reliable and fast operating light automatic rifles and mortars, such a "storm group" could probe for weaknesses and would then break-through whenever they found such an opening. Their aim was to defeat the main effective weapon of the enemy viz. machine gun by surprising the enemy and creating confusion and demoralisation in their ranks. Thus, Germans had understood machine guns, though not fully, but better than other nations. The "infiltration" did succeed to begin with in giving a tremendous speed to the march of German forces but then suddenly the attack came to a halt at 'Marne' and 'Verdun' as the Germans had outrun their supplies, for, specially as the horse-drawn artillery of 1918, and to some extent even the railways were not adequate to support such a rapid and deep advance. Also no one had foreseen that even a small number of machine guns could stop the advancement unless paid heavily in terms of casualties. No wonder the German losses during the first four months were 747,465 killed and wounded and the French had 854,000 casualties. It may be noted that Germans should have suffered a few times more losses than French, as they were on the offensive, but Germans had understood the impact of machine guns enough to bypass its fire. They followed the 'infiltration' and hence suffered significantly less casualties.

Technology Needs A Study in Depth

Technology when misunderstood or misemployed in war forces a punishingly heavy toll, just as when understood and employed properly, it brings victory faster and at a lesser cost.

France had entered WW I with the (unfortunate ?) offensive plan XVII of Gen. Joffre and when their offensive in Ardennes and Sambre areas failed miserably, French decided to defend themselves at Marne, east of Paris and at Verdun. Even by spring of 1915, both the French and British believed that artillery was the key to the deadlock created by the continuous front of trenches from the Channel to the Alps. But even after a prolonged bombardment by 1140 pieces of heavy artillery, 53 French Infantry Divisions were able to advance only 3000 yards on a 12 mile front. The Germans simply weathered the storm of shells from the enemy's guns, in deep dugouts until the barrage lifted, then quickly manned the machine guns to shoot the attackers trying to advance under the cover of artillery across the no-man's-land. Despite this experience, till early 1917, it was still thought that by increasing the density of artillery, the infantry could make headway behind a rolling barrage timed to the advance. But even a ratio of 'a gun to six yards' at the front was not enough, and German machine guns continued to reap their deadly harvest. Whilst, on the French side, merely two machine guns (French Hotchkiss) fired 75000 rounds in the famous successful defence of Verdun.

All this is related here, mainly to indicate that not only does technology have impact on the conduct of a war, on its strategy and tactics and its outcome, but it is also important to foresee the type of influence that a technology would have. Despite the experiences of appalling losses in the Crimean, the Franco-Prussian, the Boer, the Russo-Turkish and the Russo- Japanese wars in the recent past, European Generals of WW I era were so obsessed with 'offence' that they concluded that machine guns were offensive weapons rather than defensive, despite being heavy (45-70 kg). They clearly saw it as a weapon favouring offence perhaps, because they badly wanted an effective weapon for offence. Even as late as 1915, the British Commander Sir Douglas Haig asserted that the machine gun was a much overrated weapon and agreed that the French scale of 2 machine guns to a battalion was more than sufficient. In a way, Sir Doughlas was correct (?) as he was evaluating machine gun as an offensive weapon; that it could be an effective defensive weapon, probably never crossed his mind. If these technological factors could have been seen in correct perspective, without being wrongly influenced by prejudiced 'idée fixe', viz. 'offence is the best strategy', the course of WW I would have been different and the human losses would certainly have been dramatically lesser.It is this mis-judgement on the role of machine guns that prevented WW I from becoming a successful hi-tech war.

Despite this misunderstanding on the role of machine guns, other technologies did play important roles in the WW I, and electric telegraphy is another example. In 1844, the first Morse-Electric -Telegraph system was used between Baltimore and Washington. When in 1854 British and French military engineers used the same system in the Crimean War to connect London and Paris with Bulgaria, a revolutionary new era in the command and control of war had dawned: The forces can be controlled, the progress can be monitored, and orders passed from 'anywhere to anywhere'. Even, as early as 1918, though at the end of WW I, the scope and activity of military signals had increased so much that military telecommunication engineering became a separate corps in most of the armies.

Air Power in WW I

Contribution of technology in WW I cannot be fully estimated without considering the role of air power, however small. Although the role of air power, merely because of its infancy, remained marginal, yet its limited use demonstrated its great potential. At the outbreak of war in August 1914, Germany had the largest air force, almost equal to the total strength of both Britain and France. Despite initial mistakes, Zeppelins proved successful in both bombing and reconnaissance, but the airships could not survive even early stages of the war due to their vulnerability to almost any attack.

At that time, military leaders believed that heavier than air flying machines were not useful for anything but reconnaissance. The fact that the advance of German armies was held at the River Marne, because inter alia, complete and accurate information on enemy dispositions could be furnished by aerial reconnaissance. It was the success of air reconnaissance and artillery spotting that led to the idea that enemy aircrafts should be denied freedom to fly in the relevant airspace which, later on, evolved into the concept of 'air superiority.' Thus the idea of installing a gun in a military aircraft was pioneered by Britain, France and Germany but lack of reliability, and other problems prevented its proper development in Britain and France. Germany, having produced more successful versions first, enjoyed air superiority over the Western Front till May 1916, after which the Allies got back the air superiority as they had a better designed aircraft with machine guns which fired from front also.

Bomber aircrafts were developed and used successfully by Britain from November 1914 onwards, and other countries followed suit . It was the threat of German bombers that prompted integration of Royal Flying Corps and Royal Naval Air Service into Royal Air Force in April 1918. In its strategic bombing role with its five squadrons of night-bombers and four squadrons of day-bombers, it dropped 550 tons of bombs during just the last five months of the war. It resulted in destruction of German airfields with absolutely no loss of allied aircrafts by enemy during that period. The day-bomber British aircraft DH-4, which was fast enough to elude most of the enemy fighters in day-light, pioneered the bombing technique that relied on speed, rather than guns, for their defence. RAF had carried out a sustained bombing offensive against the German munitions industry. Although the achievements of RAF were not very significant, they were portent enough of things to come and demonstrated the 3-D flexibility of air power.

Technology in WW I

To complete a general picture of the role of technology in brief, it can be mentioned that technology had produced massive railway networks, telegraphic communication systems, electronic navigation systems, electronic warfare systems, electrically controlled mines, Brennan torpedo (an underwater wire controlled guided missile with a range of 1000 yards), gun cotton with detonating fuse, TNT, shaped demolition charges, balloons, powered air ships, photogrammetry (making of maps by use of aerial photography), more reliable and faster operating rifles (and bullets) and of course machine guns etc. Technology knows how to win enemies and influence military thinking and therefore can be overlooked at one's own peril and that too not for a long period. This is amply demonstrated by the fact that before the end of WW I, but after suffering heavy losses of human life, every army was deploying machine guns at a scale of one machine gun per squad consisting of 6 to 10 soldiers and attacking, while paying due respect to defences provided by machine guns.

To Attack or to Defend, Is the Question

Soon after WW I, not surprisingly, 'defence' was the major concern of most European countries. French military engineers designed and constructed the Maginot line, a highly sophisticated and unimpregnable (thought to be so) defensive system to protect their common frontier with Luxemburg and Germany which was reciprocated by German's Siegfried Line. The attacking potential of aircrafts was seen and hence anti-aircraft guns were manufactured and searchlight positions were built. For air defence of Great Britain, by 1939 a chain of radar stations on its southern coast was specially constructed, an innovation that saved the Great Britain. Gas proof air raid shelters were constructed in large numbers and important installations were camouflaged from aerial observation. Having realized the impact of rifles and machine guns in WW I, every nation having an arms industry wanted to produce an automatic rifle in which the next bullet would automatically be loaded. Sub-machine guns were also developed, e.g. the American Tommy gun, M1 sub-machine and Browning automatic rifle, the German MG-34 (air-cooled and light- weight), the Soviet Degtyarev, the British Bren etc. For defence against armoured vehicles, heavy machine guns were developed e.g. US Calibre 0.50 Browning, the Soviet 12.7 mm, the Dshk etc.

Dare Devil Pilots

In order to understand the impact of technology on WW II, let us look again at the development of military aircrafts, tanks and electronic warfare systems. Although Wright brothers had given a head start to USA by their successful first flight in 1903, USA did not foresee the military potential of aircraft as much as France, Germany, Russia and Belgium did. By the year 1914 Germans had produced the L3 Zepplin airship with a metallic body. Its engine gave it a speed of about 75 Kmph with a unique payload of five high explosive bombs (50 kg each) and 20 incendiary bombs (6.5 lbs each). Installation of machine guns in aircrafts was also experimented by Germany, France and Belgium. In Belgium, an American, Col. Lewis, began manufacturing the Lewis gun which was to become a standard aerial weapon.

Techniques for bombing and torpedo dropping from air were developed which later became a base for tactics and strategies of air battles. The evolution of aerial tactics, without an iota of doubt, demonstrates once again the validity of the axiom - 'mind over matter', specially when human mind is motivated rightly. The evolution of aircraft carriers is another example to prove this. In early 1916, the first land-planes were being flown off the 61 meter decks of pioneering carriers, hastily converted from merchant ships. Although the planes could take-off from such a deck, they had to land back on the sea and then hauled back on board by cranes. On August 2, 1917, a dare devil pilot ventured and landed, against the established practices, his aircraft on the take-off deck of HMS "Furious" while the ship was underway, thereby giving birth to the concept of true aircraft carrier. So was a reconnaissance plane converted into a bomber, against all orders, by an adventurous RNAS Commander Samson, merely through his will power.

As more than 57,000 aircraft had remained in service with the French, British and German air forces, after WW I, no government was willing to spend large sums of money on new military aircrafts. This was despite the few spectacular actions that had demonstrated the potential of air power and also despite the "prophecy" made by Italian General Douhet in 1921 that, "a war can be won by air power ...". Only when the dark clouds of war started hanging in the skies during 1934, revolutionary improvements started taking place in aircraft design. All metallic monoplane aircrafts installed with several machine guns and retractable undercarriage were produced. Similarly bombers with enclosed cockpits and internal weapons-bay with bomb loads of several thousand pounds were produced. Superchargers to boost engine power at high altitudes and closed cockpits enabled the aircrafts to fly upto an altitude of 20,000 feet. Blind flying instrument panels, bullet-proof wind shields and seats became standard equipment of aircrafts, thereby enabling them to fly reliably during night, through bad weather and through anti- aircraft fire as well. More accurate gun sights and bomb-sights were developed. Perhaps the most important and innovative of all, the pilots of RAF Fighter Command were trained to be guided toward their targets (enemy aircrafts) by ground radar operators (fighter controllers). This was a unique synergic action (similar to integration of eyes, brain and hands in our body) which made the technological wonder "Radar" an effective and formidable defence system against air attacks and is probably the most important reason for reducing the effectiveness of the far superior German Luftwaffe in the 'Battle of Britain'. Both the nations had comparable hi-tech capabilities. It was the 'slightly' superior hi-tech ('electronic superiority') capability of British scientists and technologists that enabled Britain to 'defeat' the mighty Luftwaffe in 'Battle of Britain. Developments also took place in various guided missiles but except for air-to-surface missiles, none could prove significantly effective.

Role of Tank in WW II

Another hi-tech weapon of WWI & WW II was 'tank' and it would help us in understanding the role of technology if we have a brief look at its development. Although the idea and invention of an armoured fighting vehicle is as old as vehicles themselves, the modern vehicle that was both armed and armoured was constructed first in London and then in Paris in 1902. By 1904 a turreted armoured car was also built in Austria. Proposals for tracked armoured guns, made in 1903 in France and as late as 1908 in Britain, and even later in 1911 in Germany and Austro-Hungary, were rejected by military authorities ! Time and again military authorities fail to realize the potential of a new hi-tech weapon, but soon technology proves its necessity. High mobility and offence were the notes on which WW I began with a bang which was helped by the accelerated development of armoured cars, but very soon it whimpered into an obstinate trench warfare. Trenches proved the armoured cars to be impotent, and need for tracked armoured vehicles was then felt.

The first tracked armoured vehicle was improvised in July 1915, in Britain, and the first tank "Little Willie" was constructed in September 1915. In February 1916, an order for 100 numbers of an improved version "Big Willie" was placed by the British Army. At the same time French Army placed an order for 400 tanks, but could get and use them only by April 1917. 49 British tanks were put into operation on September 15, 1916 but the number was too small to be really demonstrative. On 20 November 1917, 474 British tanks were concentrated at Cambrai, where they achieved a spectacular breakthrough. Need for better tanks was felt, as these could not be fully exploited because these tanks were slow (6.5 kmph) and had short range (30 to 65 km). It was only by 1918, 14-ton Medium 'A' tanks with a speed of 13 kmph and a range of 130 km could be produced. The French tank Renault FT, a 6-ton slower vehicle proved to be more usable. It fitted well with traditional ideas about the primacy of infantry and the French accepted the tank as a mere auxiliary to infantry. The losers in the war, the Germans, despite their superior technology, had not given significant importance to tanks!

Britain took the lead in development of tanks after WW I. The tanks - Vickers Mediums - capable of 30 kmph were used by the Royal Tank Corps to develop mobile tactics. It is ironical that Germans learnt more from this British exercise than British themselves. French also produced heavier tanks with turret mounted 75 mm guns. In 1928, experimental model of a hi-tech tank capable of 68 kmph was produced in USA, which could also run on independently suspended wheels after removal of the tracks. This enabled them to run comparatively faster on broken ground. The next improvement was to increase the thickness of armour so as to withstand full effects of anti-tank guns. By 1934, Germany had realized the value of tanks and had started production of the same but with a 37 mm to a 50 mm gun. Russians realized the value of tanks with higher calibres, and in 1939 they started producing medium tanks with a 76 mm gun.

Leader Falls Behind

Although Britain had taken a lead, but on the eve of WW II, all but 80 of the 1148 British tanks were still armed with machine guns. France had a more powerful tank force - 2677 modern tanks. Although by 1939 Russia had a massive tank force of 20,000 tanks, German tank force of 3195 tanks proved to be the most effective. Unlike all other countries which had divided their tank forces between various infantry and cavalry tank units, the Germans concentrated their tanks in Panzer divisions. It took full two years of thrashing by German Panzer divisions for Allied countries to realise and reorganise most of their tanks into similar formations.

Electronic Warfare : War With Invisible Rays

The hi-tech field of electronic warfare, which played a dominant role in WW II and in later wars, needs to be discussed. It has not been given its due place, most probably because of an aura of secrecy surrounding it and also because of its abstract nature, as it belongs to the domain of invisible electro-magnetic waves (emw). The use of radio waves, part of the spectrum of emw, for communication is well known. It was demonstrated, as early as, in Russo-Japanese War (1904) that radio communication had revolutionised the tactics of naval warfare, by taking it up from deck height (lamp signalling) to ionosphere (global signalling). It was but natural that radio communication and therefore electromagnetic spectrum itself became an important target in a war. Telegraphic communication was jammed during the Russo-Japanese war and during WW I, e.g., during a naval operation in Mediterranean, when naval HMS Glaucestor's communication to Admiralty (Naval Headquarters) in London was successfully jammed by the German battleships 'Goben' and 'Breslav' which thus avoided an interception and damage by British battle ships supported by a bigger fleet.

Radar, The Penetrating Eye

Until the invention of radar (1935), the radio had provided human beings, long distance 'vocal chords' and 'ears' only, capable of communicating to any distance. Improved radars in 1938 provided 'eyes' which were capable of seeing things at hundreds of miles. In 1939, the British erected a chain of radars along the southern coast of England which paid them rich dividends in the 'Battle of Britain'. The development of electronic warfare is better discussed, though briefly, as we discuss WW II.

Diffident France Dilli-Dallies

With the memory of two wars in the last 50 years, and learning from the ("offensive") mistake of WW I, French became very cautious and very defensive. They built very strong steel and concrete defensive structures, considered to be impregnable, along the Franco-German border, known as 'Maginot Line'. Germans also constructed its counterpart Siegfried Line (West Wall), which had more flexibility as it was a thick chain of about 3000 small and mutually supporting pill boxes, observation posts and troop shelters. For defence against tanks, it depended on rivers and lakes and upon five rows of pyramid shaped reinforced -cement concrete projections, came to be known as 'dragon's teeth'. The Germans had positioned only 23 divisions along the 'West Wall' because their army, unlike the others, was designed to be highly mobile, whereas the French had put almost 100 divisions, main part of their army's total strength, along the Maginot Line. The rest of French army was put across the Franco-Belgium border. French had learnt a lesson from the mistake that they had committed during WW I i.e. they had to pay heavily for being offensive. Therefore, this time they had followed defensive principle in their strategy, another 'idee fixe'. This was again a blunder because while they (French) remembered the casualties inflicted by entrenched machine guns on the offensive forces, they forgot the victories earned by the tanks against the same duo. They also could not see the full offensive potential of aircraft.

The number of tanks that various countries possessed till 1939 is shown below in Table A :

	Table A: Until 1939		Table B: Yearly Production					
	Tanks Light	Tanks Heavy	1939	1940	1941	1942	1943	1944
Britain	1,068*	80	969	1,399	4,841	8,611	7,476	–
US	300		–	331	4,052	24,997	29,497	17,565
Japan	2,000		462	1,023	1,024	1,165	776	342
France	2,505	172**	–	–	–	–	–	–
Germany	2,984	211	249	1,460	3,256	4,278	5,966	9,161
USSR	20,000		–	2,794	6,590	24,668	20,000	17,000

* With machine guns.

** 76 mm guns.

Seed Sprouts in Other Countries

From the Table 'A', it is clear that USSR had truly realized the value of Tanks and had an inventory which was much more than that held by rest of the countries. Amongst rest of the countries Germany was second, both in numbers and quality, and unlike any other country, it had armoured divisions which could be used to create breakthroughs, an improved form of "infiltration", where tanks replace machine guns of WW I.

It is not that military thinkers of Britain had not grasped the potentialities of tank. Maj. Gen. Fuller, known as one of the fathers of armoured warfare, saw in the armoured vehicle, a combination of firepower, extreme mobility, and armoured protection, the best answers to overcome defensive forces relying on machine guns. He saw that this was particularly suited to an insular country, protected by a strong navy and air force, with a relatively small army intended primarily for expeditionary purposes in support of allies. But, this concept found greater support in Germany and USSR, and to some extent in Charles de Gaulle. The same fate was met by another British military thinker, Capt. Liddle Hart. He called for renewed emphasis on mobility, audacity and skill, to out-think and out-flank the enemy, a strategy for a fairly small highly professional force, equipped with the latest technology.

It was a pity that the British thoughts produced a General Guderian in German Army but not in British Army where some thought that Hart was advocating a defensive strategy. It was Guderian who developed techniques for quick victories - the well known 'blitzkrieg'. Further, Hitler employed the principles of 'grand strategy' by combining diplomacy, propaganda, espionage, history, geography, economics, technology, industrial production, morale and demagoguery. It must be admitted that whilst his grand design proved surprisingly successful in the earlier period, Hitler committed many mistakes. One example of his mistake is that his grand design was also motivated by his concept of 'Aryan' being the master race.

Technology Shall Prevail

The Table 'B' demonstrates the fact that technology prevails even if the concerned nations are unable to perceive and foresee the realistic impact of the technology on the course of war.

If at the beginning of WW II, France had taken the offensive whole heartedly and attacked Germany through its common frontiers with Germany, the course of war would have been very different mainly for two reasons. Firstly Germany had only 23 divisions along the West Wall against about 100 divisions of France, better than average ratio for attack and good enough to compensate for the deficiency in quality of tanks. Secondly, Germany was concentrating its forces in northern area.

10 EUROPE: Political
1:20 000 000
UNION OF SOVIET SOCIALIST REPUBLICS
CASPIAN SEA
IRAN
SYRIA
IRAQ
BLACK SEA
CYPRUS
RUMANIA
BULGARIA
GREECE
ALBANIA
YUGOSLAVIA
HUNGARY
POLAND
WARSAW
CZECHOSLOVAKIA
GERMANY
DENMARK
NORTH SEA
BALTIC SEA
FINLAND
NORWAY
SWEDEN
UNITED KINGDOM
IRELAND
LONDON
English Channel
BAY OF BISCAY
SWITZ
FRANCE
ATLANTIC OCEAN
Arctic Circle
ICELAND
SPAIN
PORTUGAL
MADRID
Barcelona
MEDITERRANEAN SEA
ALGERIA
MOROCCO
TUNISIA
SICILY
MALTA
Tyrrhenian Sea
ISTANBUL
Str. of Gibraltar
Projection: Bonne West from Greenwich 0 East from Greenwich

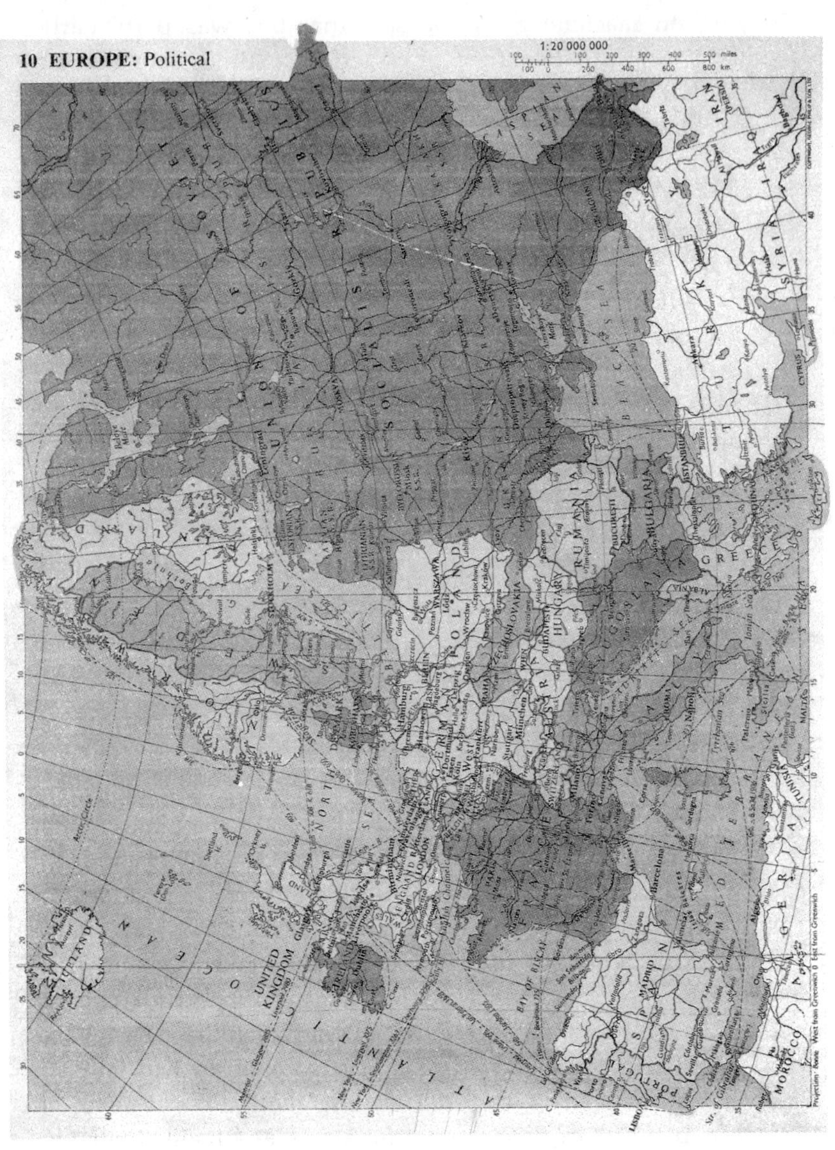

Figure 1

However, as it happened, France dilly-dallied with various ideas and then attacked when it was too late as Poland had been trampled by then (within 35 days). Germany, after winning Norway and Denmark, attacked France from across the hilly and wooded Ardennes, across Luxemburg and Southern Belgium surprising France which had expected the attack from across the northern Belgium and Holland or from their common border. This surprise was made possible by the bold use of tanks by Germans, who manoeuvred tanks through the hills and forest at a fast pace. France had deployed 41 divisions along the Maginot Line upto Luxemburg, whereas only 39 divisions along the long stretch from Luxemburg border to the English Channel. In addition, there were two French armies (12 infantry and 4 Cavalry Divisions) along with 9 British divisions amounting to about 25 divisions deployed in northern Belgium to support the Belge armies.

Germany had attacked both Holland and Belgium by dropping parachutists (a hi-tech mobility force), first to capture vulnerable points, and followed it with not only larger number of divisions but also with independent armoured corps. Through the Ardennes' hilly terrain, Germans entered by defeating the French Cavalry (horse) Divisions (in any case, an anachronism in the days of Armoured Divisions) that had come forward, and on the third day crossed the French border where they found resistance from two French armies but without anti-tank guns and anti- aircraft guns. German dive bombers and tanks shattered the French defences with ease and speed.

This short description clearly demonstrates that the proper use of hi-tech weapons secured the Germans a shocking victory over French in the early stages.

Battle of Britain, Battle of Brains

On August 8, 1940, German Air Force started massive air attacks with nearly 1500 aircrafts per day to destroy British Air Power. Against the hi-tech fighter aircrafts of Germany, British also had, though inferior in numbers but, matching aircrafts, and in addition an early warning radar chain, a set of eyes that could see hundreds of miles. Germans were always shocked to find RAF fighter aircraft waiting to pounce on them whenever they reached near their targets. It is not that Germans did not know the existence of the radar chain, but they had concluded that the radars were not functional, most probably being switched off for self protection. The reality was that Britishers had succeeded in making Germans believe that the real radar frequency was between 150-200 MHz whereas the real frequency was around 40 MHz to which the radars use to change over at the time of attack, an excellent example of electronic counter-counter measure (ECCM).

Electronic Warfare is Technology and Brains

Germans, as expected, made the radars also their bombing targets and succeeded in damaging them severely. To their utter dismay, they found that the radars were again switched-on within 3 hours of the attack. Thus, the British succeeded in convincing Germans that their apparantly successful attack on radars was not useful, as they could repair them instantly. The fact was that the radars could not be repaired so soon, but some simulators of the radars were switched-on to create the said impression.

In the first four air raids over England, the Luftwaffe lost 145 aircrafts whereas RAF lost only 88. With such a heavy rate of losses by as early as September 1940, the Germans decided to attack mainly during nights so that, despite being placed favourably by the radars, the British fighters would not be able to see and attack. The Germans could attack during nights as they had technologically solved the problem of bombing at night by means of hi-tech blind-bombing electronic aids. (For example, they had modified a civil airport landing aid 'Lorenz' for this purpose). Indeed, as early as August 13, 1940, Germans attacked at night, the factory manufacturing 'Spit Fire' fighter aircrafts (the most advanced at that time) with incredible accuracy. The British took sometime to understand and discover this highly sophisticated German electronic system known as 'X Gerat'('X' device) and later on another improved version known as

'Knickebien'. British scientists and engineers, by devoted efforts, produced a device called 'Meacon' which used to electronically deceive the 'X Gerat' of German aircrafts and guide them to drop their bombs on harmless places rather than on the real targets. After initial success of the British 'Meacon'- the electronic deception device, Germans, on November 14, 1940, again conducted night attack on Coventry, a war- industry town and Derby, the town of Rolls Royce factory, and devastated them totally. Britishers understood that Germans had employed an electronic counter-counter-measure very successfully due to which 'Meacon' had failed. British then produced another device with suitable electronic counter measure, named 'Bromide', which proved successful in later German attacks. Like this, Britishers always produced an appropriate ECM or ECCM against German electronic devices and thus foiled their night attacks.

The subject of electronic warfare would need a complete book but an introductory idea has been given here to indicate the decisive role that electronic warfare had played in the Battle of Britain. Perhaps the climax came when Britain developed an Air Defence Ground System with the help of which the British could detect the attacking aircraft many miles away, position their interceptor fighters ready to pounce upon the attackers and guide them close enough to the attackers so that even during night, with the help of an airborne radar in the interceptor aircraft, the British pilots could shoot the German attackers. Effectiveness of the British systems was so high that once they claimed to have shot 185 German aircrafts in one day. This totally demoralized Hitler and the Luftwaffe. Unfortunately for Germany, Hitler abandoned the air attacks on Britain and diverted his attacks on to eastern Europe and Russia. Although the aircraft-rich Germany had lost more than 1700 aircrafts, comparatively aircraft-poor Britain had also lost close to 1000 aircraft and was becoming desperately short of aircrafts for air battles. Had Hitler pursued, the success he was achieving with his hi-tech weapons despite the British EW capability, the course of war would have probably been different. The main course of

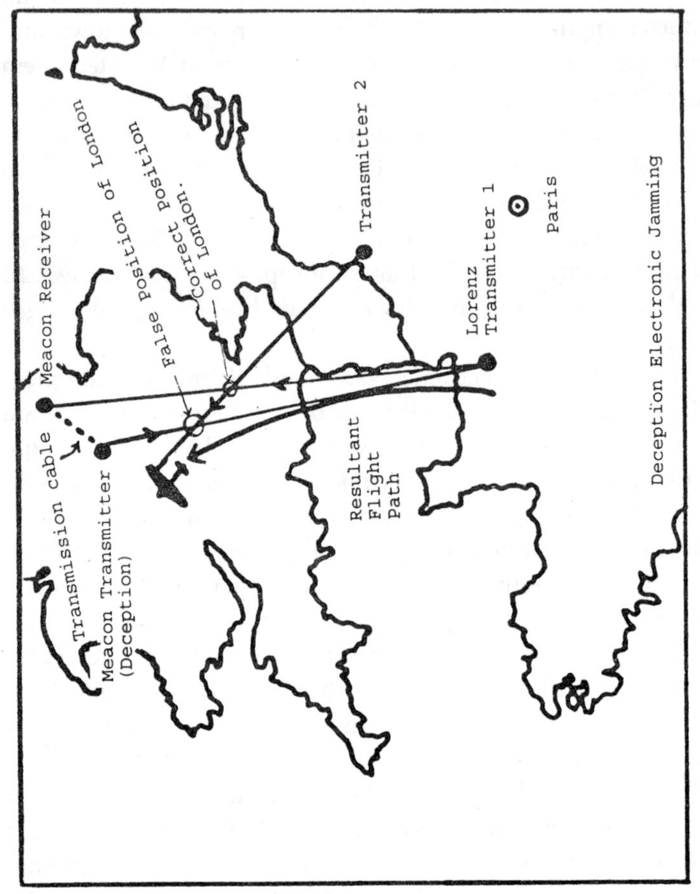

Figure 2 Lorenz and Meacon

air warfare of WW II is mainly a step by step development of aircraft as a weapon system and newer and newer techniques of electronic systems e.g. navigational aids, air defence aids, communication systems and electronic warfare. Britain won the air-war despite lesser air power, mainly because it could keep ahead in the development of electronic aids for war machinery and specially in the 'cat and mouse' game of electronic counter measures and electronic counter-counter-measures.

It is worth noting that till Germany maintained a technological lead, it could compensate for the inferiority in numbers of tanks (knowing that attacking forces have to be about 3 to 5 times stronger than those of the defending forces). Better quality of tanks and their proper deployment kept on giving advantages to Germany. When in the Battle of el Alamein, Rommel did not have petrol for his tanks, let alone the replacements for damaged and destroyed tanks, Germany had started its irrevocable retreat for defeat.

Billion Dollar Atomic Bomb

There is no doubt that tremendous improvements in war machinery took place before and during WW II but none except the billion dollar atomic bomb was so decisive as to which the victory over Japan could be attributed. All the three services of the combined Allied powers had to fight tooth-and-nail to bring the technologically advanced Germany onto its knees. The US and Britain had to take one difficult step after another to bring Japan close to defeat. The deployment of the ultimate (well, almost !) weapon, the billion dollar atom bomb, came somewhat late to save the trials and tribulations of the Allied forces, but when it came, even the obstinate Japan accepted its defeat instantly. Yet, because of its rather late apparition, neither the billion dollar bomb can credit WW II as the technological war of the century, nor even the role of armoured vehicles, as their role was misunderstood until the technology itself blasted its proper entry. It is also worth noting that there were many other vital factors which had influenced the outcome of this war, Hitler's blunders are, for example, some such factors.

Few more ideas came in sharp focus from the experiences of WW II. With the technology becoming more and more dominant and expensive, the modern wars would encompass the entire nation, its defence forces, civilian population, industry, economy as well as its morale. Production of several million soldiers with their expensive weapons for WW II was made possible by the surplus wealth (both in kind and cash) generated by technology and consequent business and trade. Clausewitz was proven right when he stated that technological development, brought about by economic, social and political changes, constantly affects tactics and strategy. It should, however, be added that technology changes, inter alia, also the structure of the armed forces. Invention of artillery razed mighty forts to ground. Invention of machine guns shot the cavalry out of existence. Air power has sunk the huge battleships. The two portions have been interchanged by mistake they should be and has converted once mighty tanks in to sitting ducks. And now the nuclear weapons are changing the world scenario. Saddam couldn't use his diabolical chemical weapons for fear of nuclear retaliation.

At the same time, Clausewitz's other concept of warfare as the collision of mass armies in which the outcome is decided by sheer numbers and will-power has been seriously challenged by the high-tech weapon systems as experienced during Kuwait War.

Arab Israeli war of June 1982 certainly deserves a. brief mention here as it was a high-tech war, though not anywhere near the Gulf War 91 in its scale and scope. On June 9, 1982 when the Israeli army reached the Syrian border, they found Syrian army alert, equipped and all set for the war: Syrian air force was also equally ready. Syrian army had deployed 600 tanks,had provided Russian anti-aircraft guns for defence of important targets, along with 20 modern surface-to-air missile (SAM) batteries, protecting the not so wide a front. Israel, on its side, had deployed its early warning radar aircraft, E 2C Hawk Eye, on reconnaissance just outside the reach of Syrian SAMs and these Hawk Eyes were also provided with a suitable fighter aircraft cover. E2c Hawk-Eye was to provide the real-time information to various Israeli fighter aircrafts about the position and movement of each Syrian fighter aircraft soon after take-off. Israeli aircrafts consisted of 'F 15s'- air superiority fighters, 'F16s' - fighter bombers, 'Kafir' and 'Phantom'- ground attack aircrafts and 'Sky-Hawk' - close support fighter bomber aircraft. All its aircrafts had electronic warfare systems. They had further deployed a number of Boeing 707 aircraft which were modified for jamming of radars, SAMs and communications.

As the Israeli army advanced to attack the Syrian army, Syrian aircraft were sent to provide close support to their army. Some 60 improved MIG 21 aircraft and some other modern Russian aircraft took-off but they got a shock of their life when they reached the battle-field. They suddenly discovered that neither their radars could see anything nor could they communicate with their 'ground control stations', i.e. they were made 'blind' and 'deaf'. Similarly, the SAM operators found out that their missile guidance electronic systems were also jammed. On the other side not only Israeli aircrafts' radars and communication systems were functioning but their Hawk-Eye and the Boeing 707 were also functioning properly and effectively. Further Israel had also used Remotely Piloted Vehicles (RPVs) for electronic surveillance and photography of Syrian targets and movement of Syrian army. These RPVs were made of fibre glass (mainly) and hence were invisible to radars. Some of the RPVs were used for illuminating important targets with laser beams to enable Israeli aircrafts to aim Maverick air-to-ground missiles accurately. No wonder very soon Israeli air force had established 'air superiority'.

In only 3 days of war (9-11 June) Israeli aircrafts had shot down 79 Syrian aircrafts and damaged at least 7; destroyed 19 SAM batteries out of the 20 deployed and Israel had lost only one aircraft out of the 90 that took part in the war. This war, as can be seen from the weapon systems used, was a hi-tech war but, as already said, it was of a small scale and limited scope, but it certainly gave a definite indication of role of an invisible kind of hi-tech weapon in future wars to anyone who wanted to see.

Till 1945, atom bomb was the ultimate weapon which brought the war to an end immediately. Therefore, hope arose that such a weapon may make wars redundant. But then many wars - Korean, Vietnam, Arab-Israeli, Indo-Pak, Afghan, Iraq-Iran, Gulf War 91 etc. have been fought. The nuclear weapon merely remained a deterrent for war between the super powers, although their nuclear pile kept on rising to levels that could destroy this small world ten times and more. Nuclear deterrent remained a strategy between the two super powers till the USSR got disintegrated. With the decline of USSR as a super power, not only the importance of nuclear power got reduced as a deterrent, major advanced countries have rediscovered 'trade' as the most powerful weapon for domination. Trade is a weapon of the strong. May we ask, 'Is trade continuation of war with other means'? Then certainly it is more dangerous than war with weapons. In trade, the enemy is welcomed as a friend, war with weapons sharpens the sense of identity and pride in own-culture but trade war creates a delightful, yet dangerous, twilight zone.

After the decline of USSR as a super power and consequent changing political equations, fall of Shah of Iran, the 8 year long Iraq-Iran war, 'Gulf War 91' took place. As described and discussed in the consecutive chapters, it has been attempted to demonstrate that it has come out as a high-tech war which can be logically termed as the most hi-tech war of twentieth century. This also conveys an assumption (analysed later) that no big war is expected in the next 5 years except trade wars. Here the term 'the most hi-tech war' needs to be explained. Today technology is advancing very fast and therefore a later day war would, more likely, use the later hi-tech weapons. If looked at simply, the latest war should be the most hi-tech war. Here, we wish to emphasize not only the use of hi-tech weapons of its period, but their use with proper understanding of their role. Thus, e.g., in WW II, if the role of tanks was understood properly and the atomic bomb was dropped on a primarily military target, instead of on innocent civilians, and also that it was dropped much earlier, whereby it could have saved more lives, we would have called WW II the hi-tech war of this century. (We now hasten to add that use of nuclear weapons is inhuman).

Another important conclusion that certainly and forcefully comes out of the experiences of WW I, WW II, Arab Israeli wars and Gulf War 91 is that technology is vitally important, indeed one of the most critically important factors to win a war. Technology, for some unknown reason has not been given a place amongst the principles of war. Perhaps it may be so because technology is as essential to war as a weapon and a soldier are. But, as has been demonstrated, technology needs a very careful monitoring, study and analysis with an aim towards its influence on fire power, strategy and tactics, not only in the present, but also in foreseeable future. Technology deserves to be a principle of war.

> "O . Rama, the powerful Indrajit, having sought blessings of Fire god, is capable of attacking his enemy whilst remaining invisible" (i.e. he can attack from such a great distance with long range weapons that he can not be seen)
>
> .... Vibheeshan, in Yudh Kand, Chapter 19, couplet 13 (Valmiki Ramayana), while explaining Ravana's military strength.

...and the big wars
That makes ambition virtue ! — (Othello) Shakespeare

Chapter 2

Genesis of the Gulf War

Very often the seeds of next war are sown during the preceding wars.[1] It has been observed that the seeds sprout rather fast. Greed, ambition, pride, injustice, retribution, fear, and hunger for power nurture it and sometimes become the prime cause of war. The roots of the Kuwait war can be traced to the rise and fall of Ottoman Empire, the first and the second World Wars, the Arab-Israeli wars and lastly the Iraq-Iran war.

Birth of Kuwait and Iraq

Kuwait city was originally founded by Aniza tribe of Bedouin which had migrated from Central Arabia. Before that Banu Khalid tribe was inhabiting that area. The word 'Kuwait' is diminutive of the Arabic word 'kut' which means a fortress. In 1756 Sheikh Sabah Bin Jabir of Aniza tribe founded the Sabah dynasty. Immediately after that, the East India Company entered into the scene. Since 1756 Turks were continuously attacking and exerting their might on Kuwait and in 1821 Kuwait became subordinate to Basra province of the Ottoman Turk Empire. In 1896 Sheikh Mubarak Al Sabah revolted against the Turks and sought British protection which was turned down. Britishers had treated Ottoman Empire as a useful force that kept Russia out of

---

1. For what can war but endless war still breed. - Milton.

the Near East. Realizing that the Empire was decaying, they reviewed their strategy. And almost 3 years after the request from Sheikh Al Sabah, on 23 January 1899, Britain signed an executive agreement for the protection of Kuwait. In 1913, before the WW I, the Anglo-Ottoman Draft Convention recognised the 1899 Anglo-Kuwait agreement and in 1914, Britain established a protectorate in Kuwait. On dismemberment of Ottoman Empire in 1920, Iraq was formed as a nation, for the first time in history, by amalgamation of Mosul, Baghdad and Basra; the mandate for which was given to Britain by the Allied Powers. Britain created a monarchy in Iraq in 1921 by installing Faisal I, son of the chief priest of Mecca, as the king.

Kurds Get A Raw Deal

From 637 AD onwards Iraq had become a geographical expression for flat lands between Baghdad and the Persian Gulf. As in the past, Iraq continued to be the hunting ground for various peoples, specially for Arabs, Abbasids, Mongols, Persians and then, from 16th century, for Turks. In 1921, in a tripartite conference at Ugair, Sir Percy Cox, who represented both Iraq and Kuwait, decided the boundaries of Saudi Arabia, Kuwait and Iraq. There was nobody from Iraq to really take care of its interests while the boundaries were being conceived. One may wonder as to who could have genuinely represented Iraq when the first king himself was imported from Saudi Arabia by Britishers who had formed Iraq, and if there was such a representative, would he have made any difference! Saudi Arabia was friendly to Britain, Kuwait was a British protectorate and the King Faizal I was totally indebted to the Britishers. In this conference Iraq got only Shatt al-Arab (mouth of rivers Dazla and Farat) as a passage to enter the Persian Gulf. The Kurd province was included in Iraq and not given to Mosul Turks thereby sowing seeds for another conflict, causing local rebellion which is persisting till today. After collapse of Ottoman empire, Allies created new states like Iraq, but divided Kurds into four states, Turkey, Iraq, Iran and Syria, with a total population of 24 millions. Allies promised Kurds autonomy as per the treaty of Sevres in 1920 but the promise was never fulfilled. Indeed, state boundaries created by Western colonisers have left many problems all over the world, but sometimes they had done justice too ! Iraq was not given Kuwait, which was restored to its original independent status, though it was part of the Turkish Basra province before war. Obviously and justifiably Sir Percy paid due regards to the longer history of Kuwait than the short period of occupation by the Turks.

Kuwait, Oil and Iraq

In 1923 the border between Kuwait and Iraq was mutually defined, and in 1932 when Iraq became fully independent, Prime Minister Nuri Al Said recognised the border as defined in the Memorandum of Iraq-Kuwait Convention an Boundaries of 1923 Boundary Agreement. In 1913, Britain secured monopolistic rights for oil exploration in Kuwait, and it was in 1938 that oil exploration was successfully carried out. Mass scale oil production could start by 1946 only. Before 1946, Kuwait was home for poor Bedouins, pearl divers and fishermen. And it is noteworthy that Iraq never claimed or asked for any portion of Kuwait till it started producing oil commercially i.e. 1946. When in 1951, based on earlier agreements, Kuwait raised the question of actual demarcation of the border with Iraq, the latter asked for Warbah Island as a price which, to say the least, was too heavy and unjustified to be acceptable to any sovereign country with a long history. Again in 1954, during negotiations for supply of water from Shatt al-Arab to Kuwait, Iraq preferred a claim of 4 km stretch of the Khaur al-Sabiya coast line, lying north of Warbah and Bubiyan. In 1956 Kuwait rejected a British suggestion to lease Warbah Island to Iraq in return for water from Shatt al-Arab. In March 1973, Iraqi forces occupied a Kuwaiti outpost at Al-Samita, an operation in which one Iraqi and two Kuwaiti soldiers were killed. Though Iraq had to withdraw due to pressure from Saudi Arabia and Iran, it continued to make such demands and Kuwait continued to refuse them. During 1980s, per capita income of Kuwaitis was 15,000 $ per year, the highest in the world. If only recent past is considered, one may like to sympathise with Iraqi's demand. But if one looks with deeper historical perspective and democratically, then one sympathizes with Kuwaitis. One may also feel that when such contradictory perspectives are present, then, who should decide which is correct ? Greed, power, politics or the people who are affected ? But what if the people who are affected are also declared to have a vested interest ? Should one really be permitted to succeed in imposing one's point of view by creating confusions and by use of brute force, however genuinely one may feel that he is right ?

Military Power Rules. O.K.

In 1969 when Iran's military strength was at its peak, the Shah of Iran took away the Eastern part of Shatt al-Arab from the then not so strong Iraq. It was naked demonstration of power and greed but at that time Iran was USA's great ally. When Iraq increased its military strength, of course, with the help of USSR, and when the Iranian mullahs were busy trying to establish their authority after dethroning the Shah, Saddam Hussein, in September 1980, attacked Iran with a bang for getting the Eastern part of Shatt al-Arab back. This war lasted eight years, the longest war of this century with several advances and retreats . About one million lives were lost and about 60 billion US Dollars (450 billion Dollars as per another estimate) were spent. When both the countries got fully exhausted and fed up with such a costly war, then under an international resolution the war was stopped. The most ironical part of this long and expensive war was that both the antagonists, after the cease fire, stood at exactly the same place geographically as before the war. Indeed, if Iraq had not gone to the war, it would have become one of the richest countries like Kuwait, in that period of eight years. But, should not a nation try to redeem its honour and land from a usurper ?. Are there peaceful methods available which would restore justice effectively?

Iraq's Economy is Shattered

Fortunately for Iraq, many countries like USA, France, Saudi

Arabia and Kuwait, out of their own political interests, helped Iraq in this war. During the war Iraq's economy was shattered, and it faced a deficit of 80 billion US Dollars. The table below shows part of the details - not including the debts.

	1981	1983	1989
	(Figures are in billion US Dollars)		
Govt. Revenue	23.9	17.1	34.42
Govt. Expenditure	36.4	30.8	57.6
Deficit	12.5	13.7	23.2

It is worth noting that in 1989 when the GNP of Iraq was about 35 billion dollars, that of Kuwait was 23 billion dollars.

It makes us wonder, of course, any person dwelling on the subject would wonder, as to what all was happening in Iraq. After a bloody, treacherous and devastating war lasting for long 8 years, any nation would want peace and time to lick its wounds, rebuild, give people a respite to start living normally. But Saddam could not wait two years, we wonder why ?

Guns Yes, Bread No

People of Iraq had celebrated their so called victory with Iran-Iraq war for fifteen days. They, then soon realised the futility of the celebration, they did not have the necessities of daily life. Saddam, who had prohibited import of goods during the war, opened the imports. But the people did not have money! The nation did not have money ! Instead of producing wealth, the nation had been bursting it for the war[1] machine. The nation had incurred a debt of about 70-80 billion US dollars. This debt had to be paid back, with interest within a reasonable time. It is not difficult to calculate (approximately) and conclude that it was an extremely difficult situation. The Iraqi revenue was 34 billion US dollars in 1989. Let us assume very optimistically that Iraq could double the revenue, to say 68

---

1. Top three arms exporter to the Middle East during 1984-88 were : Soviet Union (26.5 b$), the U.S.(16.3 b$) and France (12.7 b$). Top three importers of arms in the Middle East (1984- 88) were : Iraq (29.7 b$), Saudi Arabia (19.5 b$) and Iran (10.5 b$).

billion US $, then after deduction of the Government expenditure of say (a highly optimistic estimate) 63 billion US $, merely about 5 billion US $ would be left as surplus. Under, normal circumstances, with an interest rate of 7 to 8%, Iraq could not have paid even the interest every year. But it could have been made possible by raising the OPEC oil prices and also appealing to the kindness of the creditor nations for reduction in interest rate and for writing off of some amount etc. But would Saddam, a dictator, an ambitious leader wanting to lead the Arab world, appeal to kindness of some body! Instead he, through his deputy, demanded 10 billion dollars from various Sheikhs in return for protection that he had provided to them by fighting a war with Iran . Nobody wanted such a protector. In his mind therefore capture of Kuwait was the only possible solution. Saddam was fond of power game and knew well the power of media. He has always been generous with media persons. His allocation for gifts to media in 1988 was over 57 million US Dollars. Despite an astronomical debt burden on Iraq, he increased the allocation for gifts to media in 1989 to about 70 million US Dollars.

A Journalist is Hanged

It is quite clear that Saddam's attack on Kuwait was a grand plan. He had been using any incidence or non-incidence to project himself as a tough Arab leader who only could protect Arab interests against the Western Zionists Imperialists. Hanging, after all the necessary rituals, of British journalist Farzad Bazolt working for London observer was a drama to convince the public about his tough attitude towards the West. Bazoft had been caught outside the periphery of the military industrial complex at Al-Hillah, collecting samples of the sand there. For this he was declared to be a spy working for Israel and hanged with full travesty of justice.

Saddam Threatens

Realising the conditions and attitude of Iraq, Saudi Arabia condoned the debts, but Kuwait said it would also consider condoning provided the borders are demarcated. Iraq started accusing Kuwait of theft of oil from their Rumallah oil fields. Selfishness compounded with vindictiveness and military might do distort one's vision. Although the Iran-Iraq war had left Iraq in deep financial crisis, the Western powers helped it in rebuilding its defence forces and Iraq came out of the long war militarily more powerful. Rumallah oil-fields are indeed so large that they straddle lands of both Iraq and Kuwait and therefore legally both the countries can and do extract oil from the same oil- field. Iraq also alleged that Kuwait was extracting more oil than its authorised quota and thus depressing the oil prices to harm the interests of Iraq. After prolonged accusations and counter accusations between Iraq and Kuwait, during mid July, Saddam openly threatened to use force against any Arab oil exporting nation which continued to pump oil beyond its quota[1]. It made the OPEC countries to meet in Geneva on 26 and 27 July 90. In that meeting Iraq had put in three demands (i) The Southern portion of Rumallah oil-fields be given to Iraq (ii) Kuwait should pay them 2.4 billion US dollars as compensation for siphoning out oil from Rumallah oil-wells and (iii) Kuwait should compensate the losses that Iraq has suffered in the oil market and also for part of the damages that Iraq has suffered in the war with Iran by writing- off the debts and, in addition, pay appropriate amount in cash.

Kuwait, in its replies to the demands, simply stated that it would consider to write off the debts and also to pay compensation only if Iraq agreed to demarcate the boundaries. This was a big concession from Kuwait, as it had never agreed so in the past. Obviously Kuwait was trying to appease Iraq in order to avoid a war which appeared imminent. However, no solution could be found in the OPEC meeting.

Grand Plans

After failing to achieve his dream on all the fronts i.e. diplomatic, economic and psychological, Saddam hit upon a rather bold plan to capture Kuwait. Such a thought follows from

---

1.   The price of oil in the world market was 20.5 $/barrel in early 1990 whereas it had dropped to 13.6 $/barrel by mid 1990. A drop of 1 $/barrel in price was resulting in a loss of 1 billion $ to Iraq.

a definition of war (Clausewitz), viz. War is (merely) continuation of policies with other means. If one studies Saddam's life, one would find that his nature fits reasonably well with the nature of war itself involving the trinity of three elements viz. violence & passion, courage to accept challenges, and political ambitions. To him war thus appeared to be a valuable instrument of his 'national policy' as he had built sufficient military power to enable him to gain his political purpose in the 'most economical' way. The USSR Government, struggling with its economic revolution, perestroika and glassnost was too busy to put any pressure on any side or to help or dissuade Saddam in fulfilling his dream. He thought (rightly) , it was only USA which could come in the way of his dreams becoming a reality. To reduce the uncertainties to the minimum he decided to explore the American thinking. To get the reactions of USA, on such an eventuality, he discussed the subject of Iraq-Kuwait relations with Ms. Glaspie, the ambassador of USA. What exactly transpired between them has not yet come out clearly. Iraq, on one hand, claimed that Ms. Glaspie had told Saddam that the USA wanted to maintain friendly relations with Iraq and would not like to interfere in the bilateral matters of Arab countries. On the other hand, Ms. Glaspie claimed later that she had clearly stated that the USA would not tolerate an attack on Kuwait. The State Department of the U.S. did not repudiate the transcript issued by Iraq. Also Glaspie's 'MEMCOM' (a memorandum of communication despatched by the Ambassadors to Foreign Department) was not made public. Interestingly, George Bush, on 28 July 90 (perhaps, too late), warned Iraq not to attack Kuwait. It is surprising that the USA, possessing the most powerful intelligence resources, came to know only on 28 July, 90, about Iraq's designs on Kuwait. The US intelligence saw the military movement merely (!) as a threatening gesture to extract some of its demands. A forceful and timely warning by the USA would, most probably, have deterred Iraq from such a risky adventure.

Incidences Foretell

Four incidences that took place in this period are worth our attention. As a consequence to Iraq's threats to Kuwait, on 27 July 90 the US Senate passed a resolution to stop the agriculture debts to Iraq. The Deputy Spokesman of the US Home Department, Richard Busher, expressed his concern over the behaviour of Iraq but at the same time, said that such an action from USA would not help in meeting the desired goal.

During the same period, the law and order situation in Trinidad had worsened so much that some Muslim terrorists had taken the Prime Minister Arthur Robinson a hostage at gun point. On 30 July 90, while giving his reactions, Richard Busher stated that the US Government would not like to interfere in the internal matters of Trinidad as they themselves would like to resolve this issue. The third event, related to Ms Glaspie, has already been mentioned.

Saddam's Size

The fourth incidence took place in the House of Representative Foreign Affairs Committee, just two days prior to Iraq's attack on Kuwait. In this meeting the Democrat Le Hamilton had asked, "How prepared is the US to assist the friends in the Gulf countries ?" Assistant Secretary of State, John Kelley had replied, "It is clear that we have no defence pact with any of the Gulf States. We are prepared to assist in defending and protecting freedom of friendly countries of that area. We have never interfered in the boundary disputes or internal discussions of OPEC countries. But we have always assisted and stressed on the peaceful ways of settlement of disputes." One can not conclude definitely that the USA was tempting Saddam to go to war, but all these incidences create such an impression. This impression is strengthened because the USA had started realising that Saddam was becoming anti-West and was wanting to flex his strong muscles to take over the Arab leadership. Domination of Middle East by an anti-West Saddam would be harmful to the interests of the West. The USA must have realised that by helping Saddam during Iraq—Iran war, they had created a Frankenstein. The USA, therefore, must have given a thought to cut Saddam to size.

Vietnam Syndrome

Vietnam couldn't dominant have been a cause of this war, it still was one of the motivating forces. Military frustration and defeat by a small third world country was haunting the psyche of Americans like a nightmare. In a Survey by Gannett Fountain Media Centre it was found that the word Vietnam appeared 7299 times, three times as often as the runner up 'human shield'. Saddam Hussein wanted to repeat Vietnam and USA had learnt many important lessons.

2nd August, 1990

As a result of these four incidences, and also the fact that the USA and other countries had helped Iraq in the Iraq-Iran war, Saddam must have concluded that his golden chance to fulfill the age old dream of Iraq had arrived. He appears to have concluded that so long as the USA was convinced that Iraq would not harm their allies' interests, the USA would not involve itself in a war over Kuwait. On 2 August 90, giving three reasons, Saddam attacked and occupied his small neighbour[1], a sovereign nation. The first reason he gave was that Kuwait had depressed the oil prices in the world market. The second was that Kuwait had not paid compensation for the oil it had pilfered. The third was that Kuwait was illegally occupying Iraqi territory. War has, sometimes, been defined as 'organised mass violence' and this is how Saddam launched this war.

It is surprising that having known the amassing of troops by Iraq in real time, the US concluded the action as a mere threat. Even Israel had provided evidence on 20 July 1990, that Iraq had deployed offensive missiles along Kuwait and Jordan borders. At least on 23 July when the US intelligence satellites had confirmed that atleast 30,000 troops had lined up on Kuwaiti border, the US could have issued a press statement to warn Saddam. It is also surprising that Kuwaitis, despite their long experiences with wily Saddam, did not suspect this attack till the

---

1. Kuwait : Area - 18,000 sq. km., population - 2 million (Kuwaitis only 0.57 million)

   Iraq : Area - 438,000 sq. km., population -18 million (Shiite - 55% and Sunni 40%)

last moment. Of course it also speaks volumes of the cleverness of ploys that Saddam has been putting up.

A Dagger Thrust Into Arab Brotherhood

King Fahd of Saudi Arabia was finding it difficult to believe that, contrary to his recent promise, Saddam had attacked and occupied Kuwait. He was aware of Iraq's military strength and also that his own army comprising 65000 soldiers, 500 tanks and 180 fighter aircrafts etc. was no match to Iraq's. He was depending upon the understanding reached between the Arab countries that they would not attack a brother Arab country. In the recent history of Middle East, this attack was the first dagger thrust into the Arab friendship. He was clearly worried, because he was not at ease in inviting a non-muslim army on his holy land to fight a muslim nation, albeit an attacker.

Quickest Victory: Longest Misery

It is worth noting that during July end, General Schwarzkopf as the Commander-in-Chief of the Central Command had presented a scenario for an exercise, as per which one dictator had attacked a neighboring country in the Middle East to capture oil wells. Schwarzkopf had asked his commanders to give a military solution to the problem. Within a few days, what Schwarzkopf had 'imagined' only as a paper exercise, had become a historical fact. In the Defence Forces, such exercises are conducted and are treated as a routine matter, but extremely few of them become a fact of history. Saddam just marched in with one lakh modern army (5 times the strength of Kuwaiti's army) along with air and naval elements. It took only 4 hours to occupy Kuwait, probably a world record of quickest victory. All that Kuwaiti's could do was to appeal on private radios: " O Arabs, Kuwait's honour and life are being violated, rush to save it !"

Righting the Wrong Begins

This extremely unjustified, unprovoked and high-handed attack by Iraq produced very quick and strong reactions in the USA, Japan and West European countries and rest of the world as well. Persian Gulf is the source of about 65% of oil consumption of Japan (although only 13% of Japan's need of oil comes from Iraq and Kuwait), of 10 to 35% of oil consumption of West European nations and of only 11% of the oil consumption of the USA. But it is a very important element in the USA's power matrix. No country had doubts on the bullying and unethical nature of the attack, but to find appropriate courses of action, to set the things right, was also proving to be difficult. Even in the USA there was a powerful group advocating that this incidence would have negligible impact on the USA and therefore there was no necessity for the USA to entangle themselves in this war. The party in power thought that if Saddam was not halted immediately, the very next day he could drive his tanks on the coastal highway and capture the Saudi's oil wells on the east coast. Then the dictator would be master of 40% of the oil resources of the world and it should be remembered that he also has fourth largest army with modern weapons. A dictator gains power to fuel his ambitions and realise them. After a detailed analysis of the situation, USA decided to put the things in order, well, a 'New world order'.

Desert Shield is Created

The whole world condemned this uncivilised attack on a small neighbour (Kuwait) by a powerful and big country (Iraq). The same day (2 August 90) the issue was raised in the UN Security Council and with surprising,though desirable,speed the Resolution 660 was passed, which demanded that Iraq should immediately vacate the possession of Kuwait without any pre-conditions. When this resolution had no effect, on 6 August 90, the UN imposed economic sanctions and trade embargo on Iraq. The naval forces of USA and Britain started imposing this embargo effectively for what is the worth of an embargo in this advanced and civilised world, unless enforced by military might! But Saddam, on 8 August 90, instead of vacating the aggression, declared Kuwait as its nineteenth province. This declaration may remind students of war, Hitler's declaration of Austria as a province of Germany, prior to the WW II. This declaration was to serve a dual purpose. It indicated to the world that he was dead serious about his actions and would not retract. Secondly, since victory consists not only in the occupation of a battlefield or an area, but in the destruction of the enemy's physical and psychic forces, Saddam aimed at destroying Kuwaiti's morale as well. Whatever impact it had on Kuwaitis, USA reacted contrary to Saddam's expectations. George Bush after being invited by Saudi Arabia, consequently, declared operation 'Desert Shield' on 8 August 90. The US sent their fighter and attack aircrafts F-15s and F-16s to Saudi Arabia. Naval forces were also despatched by USA and Britain.

Drive the Aggressor Out : UN Mandate

When Saddam saw a severe reaction to his occupation, he, in order to gain at least some Arab sympathy, conditioned his evacuation of Kuwait with that of Palestine by Israel. USSR and France made attempts to resolve the issue through the auspices of the UN but did not succeed because of Saddam's insistence on connecting the issue with Palestine. When the economic and trade embargo failed to achieve the objective, the UN Security Council, on 29 November 90, passed the Resolution 678 which empowered the Multi National Forces (MNF) under the leadership of the US to use any means to drive Iraq out of Kuwait and establish peace and stability in the region. Iraq was given time to vacate Kuwait by 15 January 91. Iraq retaliated by threatening to use chemical weapons and putting fire to the Kuwaiti oil wells in order to defend its position. This was the first case in the history of war when 'destruction of environment' with no significant strategic or tactical (military) objectives was being used as a 'threat' to win a war, although environment is always the first casualty in any modern war, be it Hiroshima or Vietnam.

Most probably, Saddam had concluded in his mind, based on the incidences quoted that the US would not attack Iraq. He must have further thought that for a very small gain, (The US imports only about 10% of its oil requirements from the Persian Gulf), USA would not, having learnt from Vietnam, risk the valuable lives of American soldiers by pitching them against the well trained Iraqi army in an extremely hostile environment of desert. Saddam, perhaps, also thought that the US was only concerned about his friend Saudi Arabia and as per operation 'Desert Shield', wanted to ensure its defence. Saddam might have also believed that if he could make his intentions absolutely clear to Saudi and the US of having no designs on Saudi Arabia, the US may not get involved in the war. Thus, it appears that Saddam had not been able to understand USA's global view. He, therefore, ordered his army to take purely defensive positions along the Kuwait-Saudi border. The army had therefore dug trenches and raised bunds on the entire Kuwait-Saudi border, a 'ditch cum bund' defensive measure. Next to these 'ditch cum bunds', barbed wires and mines were then laid in a belt of about one kilometer width. He also declared that the ditches would be filled with oil and lighted if an enemy ever attempted to cross it.

Saddam Ties the Hands of His Army

Offence is a fundamental principle of war which though does not mean that one must always be on offensive, but it does mean that if situation demands one must take offensive. For an air force, switching from defence to offence is not a severe problem, but for an army it is. And Iraq's real strength was its army. The completely defensive posture of Saddam had tied his army's hands behind the back, so to say. Therefore it is very logical to believe that such an extreme measure was taken by Saddam not out of stupidity but only to convince the US and Saudi Arabia about his intentions of definitely having no designs on Saudi Arabia or its oil wells.

Attacker is Attacked

Till the end Saddam was sure that the US would not like to get involved in this Iraq-Kuwait war. When the opening air attack was launched by Multi National Forces (MNF), during early hours of 17 January 91, they were pleasantly surprised to find Iraq completely unprepared for a counter attack. During the attack on that night the city of Baghdad was lit as usual. There was no immediate retaliation except a small number of Iraqi interceptor fighter aircrafts did make brave but pale attempts to perform their duty at the pre-dawn attacks by the MNF aircrafts. And it took Iraq a full 24 hours before they could launch their surface-to-surface Scud missiles.

...and the big ambitions
That make war virtue ! — V.M.T.

Chapter 3

A Unique War

Generally, wars not only have a tremendous influence on the course of history, geography, technology and politics of warring nations, but also have a significant impact on the art and science of war itself. The war declared by the UN against Iraq, through Multi-National Forces (MNF) in January 91, under the UN Resolution 678, was unique in many respects. UN Forces in the past, except in Korea and Congo, had been playing the role of peace keeping forces, unlike this one wherein it approved the MNF to fight a full-fledged war and establish peace and stability in the region. This was a unique incidence since WW II, when a sovereign country was annexed by the attacker and was declared its province. But unlike WW-II no serious objection was raised by the world when Austria was annexed by Hitler, the whole world condemned this brutal and naked aggression as totally immoral. In this war, the occupational forces were clearly defeated and driven away and Kuwait became a sovereign state once again. The Iraqi forces had abandoned the motto of offence, another unique feature. For the first time, it was unequivocally demonstrated that electronic warfare, high technology and air superiority are vitally essential and organic elements in a modern war. The role of UNICEF, UNDRO, UNRWA in helping 2 to 3 lakh evacuees from Kuwait and Iraq was also uniquely effective.

This war was vividly shown to the common man on his TV (CNN Channel) like a cricket match. Peter Arnet was more effective than Sanjay of Mahabharata, in that, instead of being a narrative for a single person, this war could be seen by anybody on CNN Channel on his TV. The 'show' of the war by Sanjay to the blind king Dhrritarashtra neither prevented the war from devastating the two battling armies nor proved a deterrent against future wars. Similarly, are we to expect that despite Peter Arnet's show, we would not learn a lesson on the futility of wars?

Gulf War on TV

In cosy comfort of their drawing rooms, people saw how a war is actually fought with guided missiles and laser guided bombs. They were shocked to see how Scud missiles and their improved versions, Al-Husseins and Al-Abbases, were launched at civilians in Israel and Saudi Arabia, 300 to 900 km away, and how these Scud missiles, instead of damaging any military target, wounded, killed and terrorized innocent civilians. Pictures, though routinely taken by fighter pilots of their attacks on targets, but never shown to public so far, were shown this time the next day on TV. People were stunned to see how the laser guided bombs and IR homing missiles entered fortified bunkers through their windows with a pin-point accuracy and how a thick concrete reinforced bunker was destroyed by hitting exactly at the same point with two successive laser guided bombs. People also saw the impressive fireworks when the anti-aircraft artillery of Iraqi air defence system fired at the attacking aircrafts. In the end, they also witnessed how with a lightning speed the ground forces of MNF penetrated into the Iraqi defences, how they chased the tanks of the armoured corps of Republican Guards from Nasiria to Basra and shot them virtually like sitting ducks, and how they captured the badly mauled fourth largest army in the world. However, the TV medium due to its inherent limitations, could not satisfy the curiosity of a common man on various aspects of this unique war, and many of his questions remained unanswered. Apart from the limits of TV media, despite the best efforts of CNN, the picture projected remained somewhat restricted as it was censored by both the nations-the US and Iraq. This does not reduce the importance of the revolutionary action that the CNN took and the impact thereof on the world opinion.

Questions, TV Did Not Answer

Many war analysts and experts, till as late as mid-February 91, were of the opinion that the war would take considerably long period and could not be won by air power alone despite the air superiority and high-tech weaponry that the MNF possessed. They further stated that the war would have to be fought between the two armies on the ground, the MNF would have to suffer heavy casualties in ditch-cum-bunds, fight with guns and grenades, face the tanks on the ground where the real strength of Iraq lies both in numbers (taking the minimum 1:3 ratio for the attacking forces) and in the fire power of their guns. Only after such a baptism in blood and gore, would the MNF be able to achieve illusive victory. Apart from many military experts who were making similar statements, Senator Edward Kennedy opposed George Bush and made the following statement : "....90% of the casualities would be American. Most of the military experts tell us that a war with Iraq would not be quick and decisive, it will be brutal and costly. It will take weeks, perhaps months, and will quickly turn from an air war to a ground war with thousands, perhaps tens of thousands, of American casualities...".

After seeing only air attacks day after day for weeks, and no action from army, almost every body asked could the MNF win the war with air power alone ? Or for that matter, how important was the role of high-tech electronic warfare in establishing the air supremacy of the MNF in Iraqi skies ? Has Saddam's claim on Kuwait got any real basis ? Why did Saddam act in such a naive manner ? Why couldn't this war be avoided ? Are there any ulterior motives of the USA, other than helping Kuwait and Saudi Arabia ? What are the lessons that can be learnt ? Many such questions remained unanswered. A few of these questions

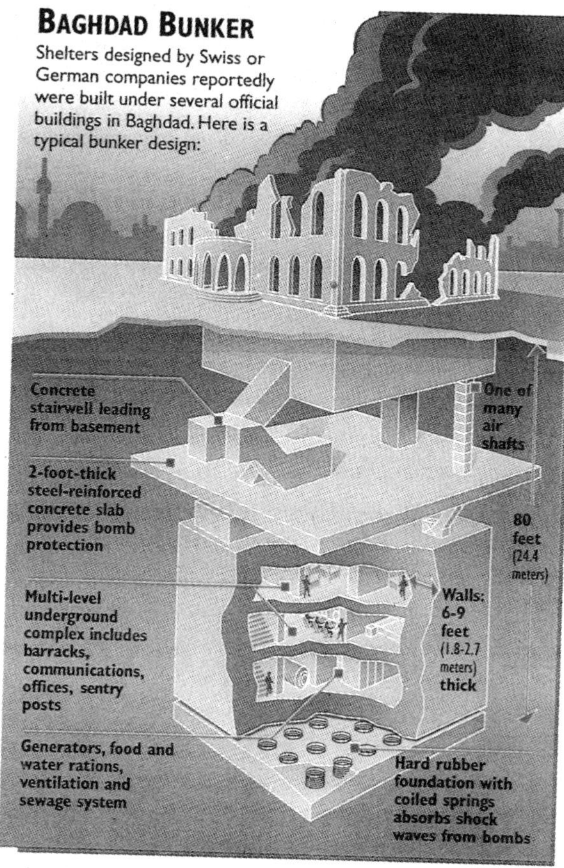

Figure 3
Baghdad Bunker

have already been answered and more would be answered in the book.

How did the MNF Achieve Air Superiority : How did the MNF establish air superiority on the very first day and air supremacy within a few days ? Whether there was a need to launch about 300 Tomahawk surface-to-surface cruise missiles (each costing US $ 800,000 approximately) and almost 4000 Maverick air- to-surface missiles (US $ 65,000 each approximately)? Why during all these attacks in the beginning, the MNF could not use the precision laser-guided bombs which cost only US $ 13,000 each, a fraction of the high-tech weapons ?. To get an insight into the answers, it may be desirable to know a few things about airpower.

Air Power

Manned flight has meant, to many, conquest of heavens by men."I touched the hand of God", said a pilot-poet after one of his supersonic flights. It is not, therefore, surprising that manned flight and therefore air power has become a rather emotional subject, in which passions, feelings and excitement abound. There is a school which visualizes that with air power death and destruction would rain from the skies; armies and navies would become redundant. Then there is another school which claims that because of the fear that anywhere nothing would be safe from air attacks, wars would be gradually substituted by negotiations. In any case, most experts agree that wars fought with dominant air power would provide quick, clean, 'mechanical' and 'impersonal' solutions.

The claims, in the early period, on behalf of air power have been, most of the times, tall and dreamy. Even ten years before the flight of Wright brothers, Major Fullerton spoke, in a conference of military experts, on revolution in the art of war that, "the chief work will be done in the air...". In World War - I, the only use for an aircraft was reconnaissance. And surprisingly the Smuts Memorandum of August 1917 described the air warfare as: "...the day may not be far off when aerial operations may become the principal operations of war, to which military and naval operations may become subordinate". In December 1919, Winston Churchill, the then Minister for Air, proved to be a visionary when he declared that "the first duty of RAF is to garrison the British Empire". As technology developed, the aircraft evolved from mere extension of eyes (e.g. artillery observation post in air) to a gun on a flying platform, to dropping of 'hell-fire' in the form of bombs. This enabled the development of a role of 'close air support' and 'interdiction'. Broadly speaking, 'close air support' is the support provided by the air power to an army stuck in a battle to enable them to face and advance against a fierce opposition; and 'interdiction' is the destruction of enemy's support structures to its army like fuel and ammunition dumps, critical bridges, communication centers etc.

As technology further developed, Edward Warner, a former professor at the MIT, in his famous essay in 1943, supporting Douhet's theory of air warfare (1921) stated that the aircraft as a weapon of war possesses the power of destroying all surface installations while remaining comparatively safe. Based on similar thinking, concept of 'strategic bombing' had already developed in which the main aim was to bomb a number of "bottleneck targets" to bring an enemy's war production to a halt, thereby rendering the enemy incapable of further resistance. But the technology then had not developed to materialise this plan and it was found that identifying the targets was one thing (although extremely difficult and yet an unreliable task), hitting them from air precisely was something else.It is also worth noting that by the time technology would develop to materialise a plan, there would be 'counters' developed to obstruct, if not prevent, the plan; but the plan would be feasible depending on one's hi-tech capability in relation to that of the enemy.

In 1945, within two years of the publication of Edward Warner's essay, the Allied powers brought the war to an end. Later on Bernard Brodie wrote, "Air power had a mighty vindication in World War - II". In World War - II, air power was used both in the form of Douhetan strategic bombers and Mitchell's tactical employment.

With the advent of nuclear bombs, nuclear war also had an impact on the concept of air power. Bernard Brodie in November 1945 proposed deterrence as the dominant concept and stated, "thus far the chief purpose of our military establishment has been to win war, from now on its chief purpose must be to avert them". Although a brave vision, alas it has not come true so far.

World War-II, despite being full of aerial warfare, had not convinced a significant portion of war theorists, some doubting Thomases, of the real potential of air power. It was when Israeli Air Force "bombed" almost the entire Egyptian Air Force in a preemptive attack in 1967 in the Six Day War, that serious thinking on the tactics and strategy of use of air power was rejuvenated. This was further given a philip by the Becaa Valley war between Syria and Israel in June 1982 when in 3 days of war, Israelies had shot down 79 Syrian aircrafts in air battles, damaged 7 aircrafts, destroyed 19 out of 20 SAM missile batteries against a loss of only one aircraft. The Vietnam war had so many problems and limitations that what it taught was "don'ts" rather than "do's", which are sometimes as useful as "do's", if not more. In 1973, the Egyptian victory over Israel brought Surface- to-Air Missiles and radar controlled anti-aircraft guns into prominence. This is what influenced Saddam deeply in formulation of his concept of 'air defence' and 'air warfare'. The technology by then had made the air defence weapons stronger than the air offence weapons, unless one possessed sufficient knowledge about enemy's electronic order of battle and sufficient support of electronic warfare system to jam the defender's electronic based weapon systems. But Saddam obviously ignored the lessons of Becaa Valley air war.

Air Defence by Iraq :

What did the oil rich Iraq do to maintain or defend its air superiority in its own skies? Air superiority means effective freedom in a given air space to use once own aircrafts and, at the same time, denying the enemy the use of that air space for use of his aircrafts. The MNF Air Force had ensured such a freedom over Saudi Arabia, Persian Gulf, and Turkey etc., but how did they establish their air superiority in Iraq's air space?

Iraq had the oldest air force in the Middle East. It also had air combat experience, having fought wars with Israel and Iran. Iraqi Air Force had also been modernized. Its pilots had been trained by Soviet, French, and Indian pilots and have been known for their courage and skill. Then how did they collapse so suddenly in air combat ?

What did Iraq expect from its air power and what were its concepts of air power. Firstly, from its air power, Iraq aimed at providing a strategic deterrent (strategic targets when destroyed, produce effective results but in a comparatively longer time frame). Its strategic resources like oil wells were valuable and their defence was extremely difficult. Therefore, as per the well known principle 'offence is the best form of defence', Iraq wanted its air force to be able to attack strategic resources like oil wells or other equally important targets of its adversary when necessary, as a matter of deterrent. However, for strategic deterrence, Iraq did not depend upon its air force alone, it had bought long range surface-to-surface missiles (like Scud-B) from USSR and then indigenously produced their improved versions viz. Al Hussein (range 600 km) and Al Abbas (range 900 km) in large numbers. Further, to increase the effectiveness of these missiles as terror weapons, Iraq developed and produced chemical warheads.

Secondly, Iraq wanted its air force to provide close air support to its army, that is, during a war if the army is not able to push forward at a desirable speed in a battlefield or is being forced to retreat by the enemy, the strike aircrafts are expected to pound the enemy targets and thus help it's army. Ordinarily, it is expected from an air force that it would maintain air superiority over the battlefield, at least where important battles are being fought. Iraq, however, did not expect this from its air force alone, and provided sufficient surface-to-air missiles (SAMs) and anti-aircraft artillery (AAA) to its army for the purpose. Thirdly, Iraq expected its air force to maintain air superiority in its air space specially in areas not covered by SAMs etc., although most of the vulnerable points and areas were protected by sufficient number of SAMs and AAAs. Such a concept of limited air defence is known as 'air control'. For the third function, viz. tactical attack (tactical targets when destroyed produce immediate benefits), which is considered essential by any air force for a successful ground offence, Iraqi Air Force was well equipped. Fourthly and fifthly, Iraqi Air Force was required to carry out surveillance and reconnaissance respectively. In surveillance, which is a systematic and continuous activity, the information is obtained on important targets and is used for planning both the battles and the war. In reconnaissance, the information is obtained by organising missions on tactically important targets and movements. Sixthly, all air forces of the world are expected to carry out counter air operations i.e. they attack targets like airfields, oil stores, ground control interception centres, maintenance spare depots, ammunition depots etc. so that the combat aircrafts of the enemy cannot take-off or even if they take-off, they cannot fight effectively. Counter air operations produce results which may have both short-term and long-term benefits. However, in Iraq such a role also was not expected. Seventhly, interdiction is another operation, expected from air force. In this action, the attack aircrafts penetrate deeper into the enemy territory and strike their vital industries, bridges, logistics elements, railway lines, vital roads and army camps etc. One form of interdiction is battlefield interdiction in which all supporting elements of the enemy formation, engaged in the battle, are attacked. Battlefield interdiction was another expectation Iraq had from its air force. Mainly as a strategic deterrance, Iraq had aimed at building strategic interdiction capability for its air force in which those targets are attacked which are considered vital to its war industry eg. oil fields and storage dumps, arms and ammunition factories, transport factories etc. Such tasks are given to Strategic Air Commands in a well developed air force but their role has remained controversial. In Royal Air Force it was in 1936-37 that the decision to switch the emphasis from strategic bombing to air defence, and Fighter Command was taken. All such differing policies indicate that Air Power Theories are in a constant flux; and policies are determined more by technological developments and their perceived influence than by theories.

Air Power Concepts of Iraq

It is clear that the role of air power was somewhat restricted in the concept of military power of Iraq. It was again demonstrated by the fact that Iraqi Air Force did not have a primary or independent position but played a second fiddle to the army. Britishers had formed an independent air arm, RAF,in 1918 itself; and in the USA, Field Manual 100-20 (published by its War Department in July 1943) - 'Command and Employment of Air Power', a document prepared by the US Army Air Forces, asserted, "land power and air power are co-equal and independent forces : neither is an auxiliary of the other".

For having given a very limited responsibility to its Air Force for air defence, it should not be inferred that Iraq did not give due importance to the sanctity of its air space. Two main reasons for such a low-key role of Iraqi Air Force have been visualized. The first is a kind of political reason. Historically, Iraqi Army had been a powerful force in the Iraqi politics and it did not want a competitor in the form of an independent Air Force. Secondly, the Egyptian concept of air control had proved a tremendous success, especially in the early phases of Yom Kippur war against Israel. The conclusion that appears to have been drawn from this is that air control is an effective concept and since it is cheaper than establishing proper 'air superiority', through air power, it has been adopted by Iraq in principle. In order to activate this principle of 'air control', Iraq had procured and deployed a large number of SAMs and AAAs around their vulnerable targets. He also gave such air defense systems to its armoured divisions and mechanized divisions, in mobile forms, in sufficiently large numbers.

Radar, An Electronic Eye

In order to understand a modern warfare, eye of the war machine must be well understood. In any case, an electronic eye (radar) is an interesting and a wonderful invention of this century.

The word radar is an acronym of 'Radio Detection And Ranging'. Radar systems have capabilities to detect any stationary or moving object on the ground or in the air, during day and night, and determine its position in three dimensions by measuring its distance and direction; and some type of radars measure its velocity as well. However, there is only one condition that the object under surveillance should be such that it is capable of reflecting electromagnetic waves impinging upon it e.g. metallic objects like ships and aircrafts etc. With the help of an antenna, the radar radiates electro-magnetic waves (e.m. waves), (either in the continuous wave mode or in pulse mode) towards the space where the object is to be detected. Rotating directional antennas are employed to search and locate the targets in all directions. The electromagnetic pulse transmitted by the radar is reflected back by the target encountered on the path of travel and the 'echo' is received by the radar. The time between the transmission and the reception is measured and is multiplied by the velocity of the electromagnetic waves (which is a well known constant) to get the distance of the target. The direction of the target is obtained as the directional antenna is made to transmit only in one extremely narrow direction, at a particular instant of time, and the direction of the target is the same as that of the antenna at the time when the echo is received. The range and direction of the target is read from a calibrated radar scope. Some radars have the capability of measuring the speed of a moving target, like aircrafts, and some track them continuously.

It is interesting to note that immediately after the invention of radio by Marconi, an engineer, Christian Hulsmyer, had successfully demonstrated a radar like system in 1904 and applied for a patent. However, at that time none of the shipping companies had shown any interest in it, as probably it was in a very primitive stage. Later, during 1930s, the work on development of radars picked up. A radar system was developed by Rudolph Kanhold in Germany in 1934 and he measured the distance and direction of a ship, 7 miles away. Now the modern radar systems are not only capable of locating distant targets, they can also control artillery guns or guided missiles to achieve accurate aims. Thus a radar on board a ship or an aircraft acts like a smart electronic eye and has the same importance as the eyes have for we human beings.

Surface-to-Air Missiles (SAMs)

Mainly there are two types of SAMs for air defence; a medium-to-high altitude system, and a quick reaction low altitude system. The medium-to-high altitude SAMs defend air space at altitudes ranging from 50 m to 30,000 m and 15 km to 100 km distances. The low level SAMs are effective at altitudes from 10m to 12,000 m with a kill range of about 4 km to 25 km. In general, a good SAM system has a probability of kill of more than 90% i.e. if it launches, say, one hundred missiles on one hundred targets, it would successfully hit ninety of them. This is an extremely dangerous situation for any enemy attack missions. Thus, how could an aircraft attack targets defended by such a SAM system and also safely return to its base ? The high-to-medium altitude SAM systems use radars not only for detection and acquisition of target aircrafts but also for missile guidance where they can aim the target with precision during day and night under all weather conditions. Only radars can search and locate the targets from far off distances under all weather and light conditions. Most of the attack aircrafts prefer to fly at extremely low levels to escape radar detection. In order to detect low level flying aircrafts, low level radars are deployed on suitable points near the border. Since such radars have a maximum effective range of about 50 Km against such low flying targets, they are located at a separation of about 60-80 Km in the form of a network.

Electronic Order of Battle

The deployment of various electronic systems like communication centres, radars, SAMs and radar controlled AAAs, used for air defence of targets is known as 'electronic order of battle'. It could be better understood with the help of a diagram given here, wherein the deployment for air defence, is only an example and it can have many variations to meet

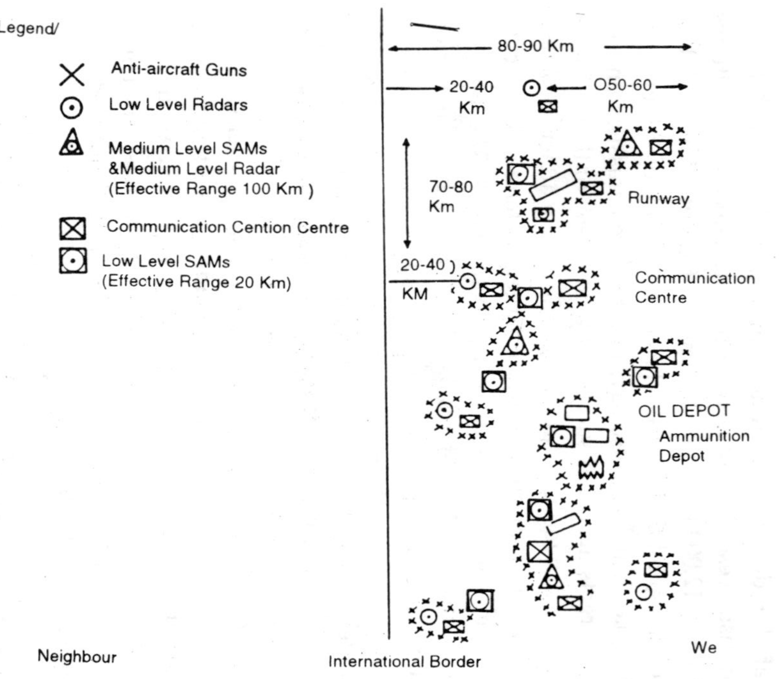

Figure 5
Electronic Order of Battle

Figure 6

different requirements. In the Iraqi air defence concepts, the density of SAMs and AAAs is larger, conforming to their concept of 'air control'.

Airborne Warning and Control System (AWACS)[1]

An important member of the air defence network is Airborne Warning And Control System (AWACS). Such a system has large transport aircraft body like that of a Boeing 707. It has a powerful surveillance radar (with moving target indication capability) which can identify and track even small aircrafts flying from the lowest to the highest altitude up to a range of 300-400 km. It has a powerful computer which keeps a record of all aircrafts flying within a radius of 400 km, gives suitable warning of enemy aircrafts, directs the aircrew of their own fighter/interceptor aircrafts to help them in intercepting the enemy aircrafts by giving the position, speed and direction of the enemy aircrafts. The AWACS has necessary communication systems which are highly jam resistant. Special receivers for listening to enemy's radio radar transmissions, which are of both long term (strategic) and short term (tactical) importance, are installed on an AWACS aircraft. These electronic receivers help in updating the 'electronic order of battle' of the enemy.

---

1. The acronym AWACS has been used here as a common noun, though the US names a particular type of aircraft as "AWACS".

Information regarding special characteristics of the enemy transmitters, their locations, radio frequencies used, modulation employed, transmitter power and in case of radars, their pulse width, pulse repetition frequency etc. are collected. Such knowledge about enemy systems helps in knowing their activities, in assessment of the capabilities of their systems, state of war preparations, and their objectives like when, how, and where the attacks are being planned. The process of collecting such information through electronic media is called electronic intelligence (ELINT) and is useful for both strategic and tactical purposes. Therefore AWACS can be a boon to the air defence system, if and only if, air superiority is maintained in its area of operation. It is worth noting that the ground based systems can detect a low flying aircraft, within a radius of about 50 km, whereas an AWACS can detect and track aircrafts, flying at any level, up to a distance of 300-400 km and control its fighter aircrafts. Iraq also had two AWACS type of aircraft. Iraq did not have special ELINT aircrafts and satellites, which were generously used with extreme effectiveness by the Multi National Forces. But, Iraq could not use its "AWACS" type of aircraft because it had lost air superiority from the day one.

Interceptor Aircraft

The attacking aircraft can safely penetrate into the enemy territory by flying at an altitude which is lower than that a radar can see (e.g. if the radar's low level capability is 50 m, the aircraft can fly at an altitude of 30 m and remain undetected) or fly behind a shadow zone (hillock or forest) even at a higher altitude. Alternatively, the aircraft can also penetrate at an altitude higher than the maximum altitude capability of the radar (if radar can see up to a maximum height of 16,000 m, an aircraft may fly, if it can, at an altitude of say 18,000 m and penetrate undetected). But, by and large the attacking aircrafts of a well equipped air force prefer to penetrate the enemy territory at a comfortable height by jamming the enemy radars as attacking a ground target accurately from such extremely low heights is not reliable. To challange the enemy's attack aircrafts, interceptor aircrafts are deployed thereby either maintaining air superiority or achieving the air control. The interceptor aircrafts become highly useful not only because of their flexibility, but also as their radars are comparatively (compared to SAM radars) more difficult to jam; and also if jammed, their pilots can locate, in the worst case, though with difficulty and at considerable peril, the enemy aircrafts visually and intercept. Such a visual contact would be helpful only if the enemy's radars have also been jammed. Iraq, in its armoury, had some of the world's best and latest air superiority and interceptor aircrafts like MIG-29 and Mirage F-1. They also had reasonably modern air superiority aircrafts like MIG-25 and MIG-21. Most of these aircrafts had the capability to fight during bad weather conditions and during night.

Ground Controlled Interception

For an air-defence system to be effective, it must have the capability to function during any weather and at any time of the day and night. Radars, in general, have this capability and hence radar guided SAMs as well as the fighter aircrafts having air interception (AI) radars on board have such capabilities but the latter also take the help of ground controlled interception (GCI) radars. In the beginning of an air combat, only the GCI radar guides its interceptor aircraft to the attacking enemy aircrafts till the on-board AI radar of the interceptor is able to see the enemy aircraft and track it. Only then the pilot can manoeuvre and fire a shell or a missile at the enemy aircraft. Such a positioning of the interceptor aircraft near the attacking aircraft is called 'ground controlled interception' (GCI). The interceptor aircrafts require such a help during day time also otherwise there is always a possibility of safe escape by the enemy aircraft. However, if need be, the intercepting aircraft may also try to intercept the enemy aircrafts without the help from GCI. Iraq had both the GCI facilities and night fighter aircrafts, even then they did not intercept the MNF attack aircrafts during night. Why ? As already stated above, it is not that Iraq did not possess well trained[1] air force, nor that they lacked the requisite courageous pilots and resources to retaliate. Then why ?. Perhaps this is another unique feature of this war which would be discussed later.

One may or may not doubt the possibility that the US wanted to cut Saddam to size; this is a fact that 2 August 90 onwards and till 15 January 91, there were considerable diplomatic activities inspired and guided by the US to make ·Saddam vacate his aggression in Kuwait. At the same time, it is also a fact that like a wise nation, whilst the diplomatic efforts were being made to avoid a war, full efforts were being made first by the US and then by the MNF to prepare for a 'blitzkrieg'.

More Human Face of the UN

Apart from 12 resolutions[2] that the UN Security Council passed, the role of the other UN institutions viz. UNICEF, UNRWA (UN Relief and Works Agency) for Palestinian Refugees in the Near East, UNDRO (UN Disaster Relief Organisation), and UNHCR (UN High Commissioner for Refugees) was commendable. First on 29 August 1990 an appeal for $ 585,000 was made by the UNICEF to help the evacuees, mostly passing through Jordan. UNICEF also released immediately $ 100,000 from their emergency reserve fund. A meeting was called by the UNDRO in which Japan pledged to donate $ 38 million, Denmark $ 5 million, Sweden and USA $ 3 million each, Norway and Switzerland $ 1.5 million each, and

---

1.   Col. Viktor Patzalyuk, the former Soviet military attache to Iraq has stated (Janes Defence Weekly, February 19, 1994) that "most of the Iraqi pilots were not well trained." It appears that he has made that statement to save the reputation of Soviet aircrafts MIG-29, SU-24, and SU-25 etc. Which country would buy such expensive aircrafts and not train its crew properly ! As per our knowledge Iraqi pilots are courageous.

2. 12 KEY UN RESOLUTIONS ADOPTED AGAINST IRAQ

660 (Aug, 2)- Condemned Iraq's invasion of Kuwait. Demanded Baghdad withdraw.
661 (Aug, 6)- Imposed sanctions on all trade to and from Iraq except for medicine and in humanitariar circumstances, foodstuffs.   **Contd.**

Finland and the UK $ 1 million each, totalling to $ 54 millions. In November 1990, Director General of the International Labour Organisation (ILO) appealed for help to countries affected by the return of migrant workers because of the Gulf War 91. UNDP, Canada and Switzerland pledged a total of 1 million dollars. Further when millions of barrels of oil was spilling on the Persian Gulf, the International Maritime Organisation (IMO) helped the Saudi Arabia to clear up the oil spill.

> "Nothing except a battle lost can be
> half so melancholy as a battle won."
>
> — Duke of Wellington.

Contd.

662 (Aug, 9)- Declared null and void Iraq's annexation of Kuwait.

664 (Aug,18)- Demanded Iraq allow foreign nationals to leave Iraq and Kuwait and rescind its order to close diplomatic missions in Kuwait.

665 (Aug,25)- Permitted use of limited naval force to ensure compliances with economic sanctions, including the right to inspect cargos.

666 (Sept,13)-Approved food shipments to Iraq and Kuwait for humanitarian purposes, if distributed by international groups.

667 (Sept,16)-Condemned raids by Iraqi troops on French and other diplomatic missions in occupied Kuwait.

669 (Sept,24)-Entrusted sanctions committee to evaluate request for assistance from countries suffering due to embargo.

670 (Sept,25)-Prohibited non-humanitarian air traffic into Iraq and occupied Kuwait.

674 (Oct, 29)-Asked states to document financial losses and human rights violations resulting from the invasion.

677 (Nov, 28)-Asked the UN Secretary-General to safeguard a smuggled copy of Kuwait's pre-invasion population register.

678 (Nov, 29)-Authorised states "to use all necessary means" against Iraq unless it withdrew from Kuwait on or by Jan 15.

"Just as military is most essential for defence of a country, and therefore its progress and prosperity, capability in electronic warfare is most essential for defence of the electronic systems of the military, and therefore its survival and victory.".   .... V.M.T.

Chapter 4

Electronic Warfare: Systems and Techniques

Known to a common man are the three dimensions of war, ie. war on land, air and sea. Fourth dimension viz.'electronic warfare (EW)' is not only new for the common man but is a covert area even for the soldiers. The development of the latest EW techniques and the related systems are invariably kept closely guarded secret, specially their exact specifications and their capabilities. The subject of electronic warfare is doubly technical as it encompasses the knowledge of electronic engineering as well as military operations. One of the reasons for the success of EW battle in 'Battle of Britain' was a close cooperation between the scientists and the soldiers. As almost all the systems of war e.g. weapons, fire control, navigation, guidance, surveillance and information management use electronics for their effective, efficient and reliable functioning, the outcome of a modern war essentially depends on electronic superiority to the highest degree. Just as military is most essential for defence of a country and therefore its progress and prosperity, capability in electronic warfare is essential for defence of the electronic systems of the military, and therefore its survival and victory.

Military electronics is a field which is developing very fast amongst the developed nations and not so fast in the developing nations. Electromagnetic waves have been used in military systems for navigation, reconnaissance, surveillance, communications, aiming, command & control, weapon guidance, platform (e.g. aircraft) and weapon control. Such systems have enabled their users to get, in comparative safety, accurate and quick results under adverse conditions of weather or light. Some aspects of electronic warfare relevant to the Gulf War would be discussed in this chapter. Aim of such a discussion is to introduce the mystifying subject to the reader, rather than give technical details.

Iraqi Interceptor Aircrafts Flew But Did Not Fight

Now, let us look at the mystery of the fact that even in the face of deadly pressure from the MNF, the Iraqi interceptor aircrafts did not put up a worthwhile fight although they could fly. Although Iraqi Air Force did not have a match for the American stealth bomber aircraft F-117A, they did have aircrafts to match the air superiority and fighter/bomber aircrafts such as US F-15, F-16 or F/A-18, albeit smaller in number. Even then Iraqi aircrafts hardly resisted the MNF air attacks and that too only for first 4 to 5 days. Some specialists feel that the Iraqi Air Force was not entirely in agreement with Saddam's views and hence did not put all their might. We do not subscribe to this view, though fact remains that more than 100 Iraqi fighter aircrafts ran away to Iran during the war. Some specialists further say that if the Iraqi aircrafts could fly they could have definitely fought, but did not do so. The question remains why? To understand this complex situation we have to know, inter alia, how an air defence system functions and how do aircrafts fight battles in the air.

It is a well known adage that fore-warned is fore-armed. This is especially true when one is dealing with aircrafts which fly hundreds of meters every second. Even during World War II, success in the air battles mainly depended on early warning radars. In the vast air space if the pilot of an interceptor aircraft himself has to detect and locate a small attacking aircraft, the possibility of his being able to search the enemy aircraft well in time is rather thin. Generally, he may detect only when the attacker aircrafts reach very near the target. Most of the time he succeeds in locating them when the enemy bombers have already bombed the designated targets and are returning having completed the mission. Then the bombers can run away leaving their escort fighter aircrafts to deal with the interceptors. If an early warning radar can provide advance information about arrival and position of the attackers, the interceptor. pilots can easily locate them and destroy them from an advantageous position. In addition, if a 'fighter controller' (a specialist, sitting on the radar scope) from a Ground Control Interception (GCI) radar can also provide guidance, the interception would be relatively easier and more effective task.

Most of the early warning radars are two dimensional radars i.e. they indicate only the distance and the direction of a target, but not its height. The GCI radar is a three dimensional radar as it has to give its pilot the information regarding the height of the enemy aircrafts in addition to its distance and direction in azimuth, and position him on top of the enemy so that he can attack the enemy aircraft from an advantageous position.

From the small number of interceptions and the way they were carried out by Iraq, it can be concluded that their GCI and other radars were not effective. It is possible only if either the GCI radars were destroyed (as the important radars were destroyed in the first few waves of the MNF attacks) or the radars (including GCI radars) were victims of electronic counter measures employed by the MNF.

Electronic Counter Measure (ECM)

The fight is always between 'light' and 'darkness', and between 'information' and 'noise'. Electronic counter measures submerge the information that enemy wants to seek either in noise or replace it with false information. What is noise ? If you remove the antenna connector from the rear end of your TV receiver set and switch it on with high brightness, you would see innumerable dots shimmering on the TV screen and these shining dots totally lack any order. The sound that you would hear at the same time would be 'noise'. In fact what you are seeing on TV and listening is 'noise'. This noise is produced by nature and by electronic components (such as used in a radio or TV) and is limited to electromagnetic (e.m.) spectrum only. The noise (e.m.) produced in nature is known as 'cosmic noise' and that produced by electronic components is known as 'thermal noise'. This noise (in e.m. spectrum) is not produced in the form of sound or shining dots, although your TV set is able to convert those electromagnetic noises into audible and visible spectrum. When a radar is jammed by e.m. noise, the radar screen also shimmers like the TV screen does without an antenna; and the disorder, the density and brightness of dots depend upon the quality and power of the jammer. The more powerful and random the jamming is, the more intense and random would be the noise produced on the victim's radar scope. Under such a condition, if there is an aircraft in the sky within the range of the victim radar, it would be extremely difficult for the radar operator to detect that aircraft, as its blips would get merged with the other numerous noise blips on the radar scope i.e. in a way, the radar gets 'dazzled' or 'blinded' by the intense jamming. Such an activity when carried out deliberately to hinder the functioning of enemy's electronic system, e.g. a radar, is called 'electronic counter measure' (ECM) or jamming.

Jamming and its Types

When a radar is jammed, it either gets blinded or deceived. Jamming is mainly of two kinds :-
1. Noise jamming.
2. Deception jamming.

As an example, if two persons are talking to each other and if someone nearby starts beating drums so that they cannot hear each other, such an act is called noise jamming. In case of radio communication the noise would have to be in the form of electromagnetic (e.m.) waves. Electromagnetic noise is produced, for example, by an old fashioned door bell, which noise one can hear on a radio or see and hear on a TV set. Lightning, in addition to e.m. noise, also produces audible noise called thunder. The e.m. noise as well as the thunder is produced because of discharge of the static electricity generated by friction between clouds. Such an e.m. noise can also be produced by electronic systems and when it is used to disturb the enemy's communication and/or radars, it is known as noise jamming. Just as one cannot listen to a programme on a radio when lightning takes place, one would not be able to listen to any message being transmitted because of heavy noise on its radio receiver, if jamming is powerful enough.

Noise Jamming : On a radar scope, one sees aircrafts as only blips (bright spots) which are generated by the e.m. pulses transmitted and then reflected by aircrafts and then received by the radar receiver. A radar can be jammed by transmitting e.m. noise on the e.m. frequency at which the radar is working. A radar scope is circular in appearance and sometimes rectangular like TV screen. TV shows pictures but radar indicates presence of objects like aircrafts by means of bright dots known as blips and not by a miniature picture of the aircraft. If there is no aircraft within its range, the scope looks like a night sky where some blips are occasionally visible here and there randomly. These randomly occurring blips are produced by nature (cosmic noise or thermal noise). As soon as an aircraft arrives within the range of a radar, a blip would appear on the screen at that distance (as per a specific scale) and in the direction of the aircraft and this blip would not dance randomly, like a noise blip but would travel as per the aircraft. The radar operator of a two dimensional radar not only sees direction and distance of an aircraft, despite cosmic and thermal noise blips, but can also roughly estimate the speed of the aircraft. The whole purpose of noise jamming is to deny this information of aircraft by submerging it in noise.

Deception Jamming - The way electromagnetic noise is produced electronically at radar frequencies, one can also produce one or more blips similar to that reflected by an aircraft in flight, by using special electronic devices. When our aircraft is entering the enemy's radar range, by means of a special device, we electronically create a moving blip on his radar screen, ahead of the blip of our aircraft (generated by reflection of the radar pulses from the aircraft), and move it with either the same or with a higher speed of the aircraft. The radar operator is likely to take the artificially produced blip as that of a real aircraft. In such a situation the enemy would first launch an attack on the 'artificial aircraft', as its blip is ahead of the real blip of our attacking aircraft. If the enemy has SAM and the blip is within its range, it will launch the missiles in the direction of the false aircraft, in the sky, where there is absolutely no aircraft. The missile will thus go waste. In such a situation he will fire second and even third missile and by that time, under such a confused atmosphere, the attackers may successfully attack the targets with bombs, missiles or rockets and return safely. Such inexpensive deceptive action by electronic means may result in no losses to the side using deception jamming and heavy losses to the enemy and is called 'electronic deception jamming'. Thus in successful deception jamming, enemy does not even realise that he is being jammed, whereas in case of noise jamming, most of the time, enemy realises that he is being jammed but he does not know the location of the jamming source.

Modes of Jamming

Active Self Protection Jamming - An attacking aircraft may carry suitable jammers which can jam the air defence radars of the enemy, to defend itself, then the jammers are known as 'Active Self Protection Jammers' (ASPJ). Every bomber aircraft, apart from the bombs and rockets to destroy the targets, would have to carry its own jammers to help him in penetrating into the enemy air defences with a greatly reduced risk. When an aircraft carries its own ASPJ, the weight of bombs and rockets would have to be reduced due to payload limitations of the aircraft. If enemy's air defence is heavy, the ASPJ would form a larger portion of the payload (weapons like bombs, rockets etc.) of the aircraft and it may, at times, become so large and the bomb load so small that the effectiveness of a bomber may become questionable. Thus there is a price to be paid for securing EW umbrella. So what is to be done when a bomber cannot carry enough payload i.e. bombs ?

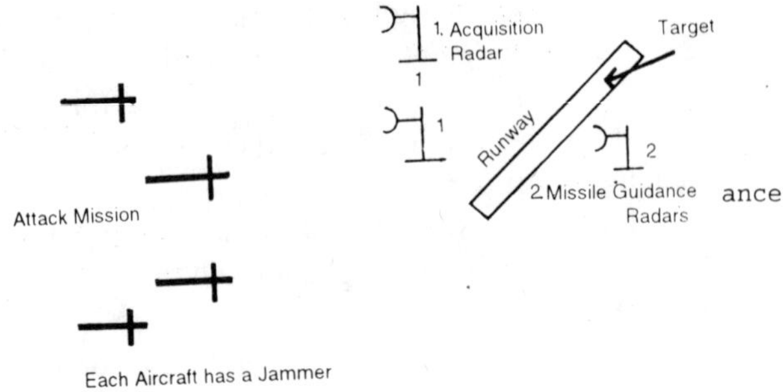

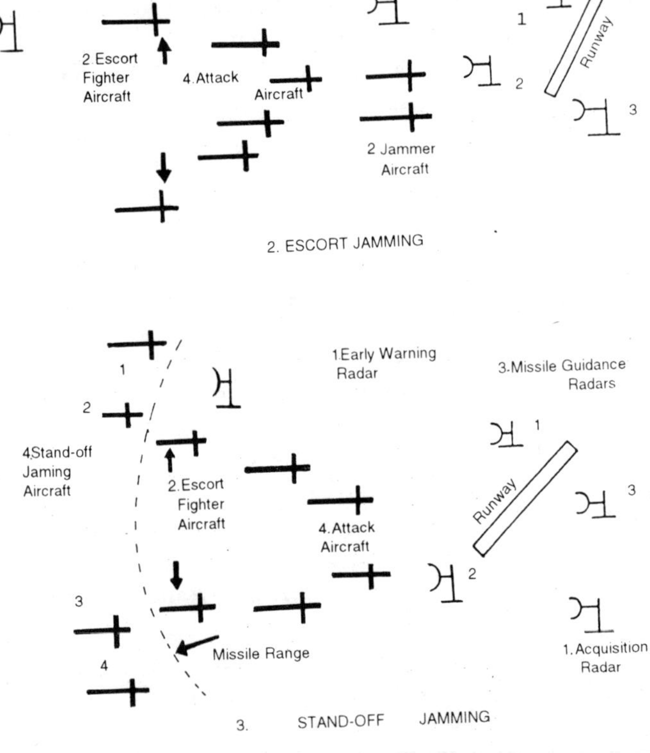

Types of Jamming (Not to Scale)

Escort Jamming : When an enemy's air defence ground environment is so strong as to reduce the effectiveness of a bomber, special aircrafts are converted into complete ECM aircrafts to meet such a contingency; e.g. the US aircrafts EF-111 and EA-6B have powerful radar jammers and EC-130 Compass Call has communication jammers. These powerful jammer aircrafts can escort 4 or 8 attack aircrafts into the enemy territory and jam all or most of the relevant radars and communication centres on the way, thus providing electronic safety to them. Such a jamming is known as 'escort jamming'. Even when attack aircrafts are given an EW cover by escort jammers, they may still carry an ASPJ system each. Such ASPJ's are specific to enemy's SAM or AAM (air-to-air missiles) and are therefore small and light. The escort jammer concentrates its jamming mainly on early warning, acquisition and GCI radars and also relevant communication. In such an attack mission, ECM escort aircrafts are also running a risk of being shot at. Well, there is a price to be paid for every benefit that one wishes to draw.

Stand-off Jamming - If the jammer of the escort is comparatively so powerful that from a safe distance, it can jam the victim radar effectively so that his attack aircrafts (though carrying a small ASPJ each) can safely penetrate the enemy's defences, the process is called stand-off jamming. This safe distance is normally more than the kill range of enemy SAMs. Thus the ECM aircrafts remain comparatively safe. The American EA-6B has such powerful jammers that it can jam most of the radars from a safe distance.

While designing an attack aircraft, efforts are made to make them light weight. A good boxer ought to have a muscular body with as little fat as possible, the same is true for an attack aircraft. Preference is given to armaments, fuel, command, control & communication system. But the jammers have become an essential part of its payload since electronically guided air-to- air and SAM missiles have entered the battlefield with deadly accuracy. The fighter aircrafts have to strike a balance between the weights of jammers and armaments. In view of this, the fighter aircrafts prefer to carry (ASPJ) jammers only for guided

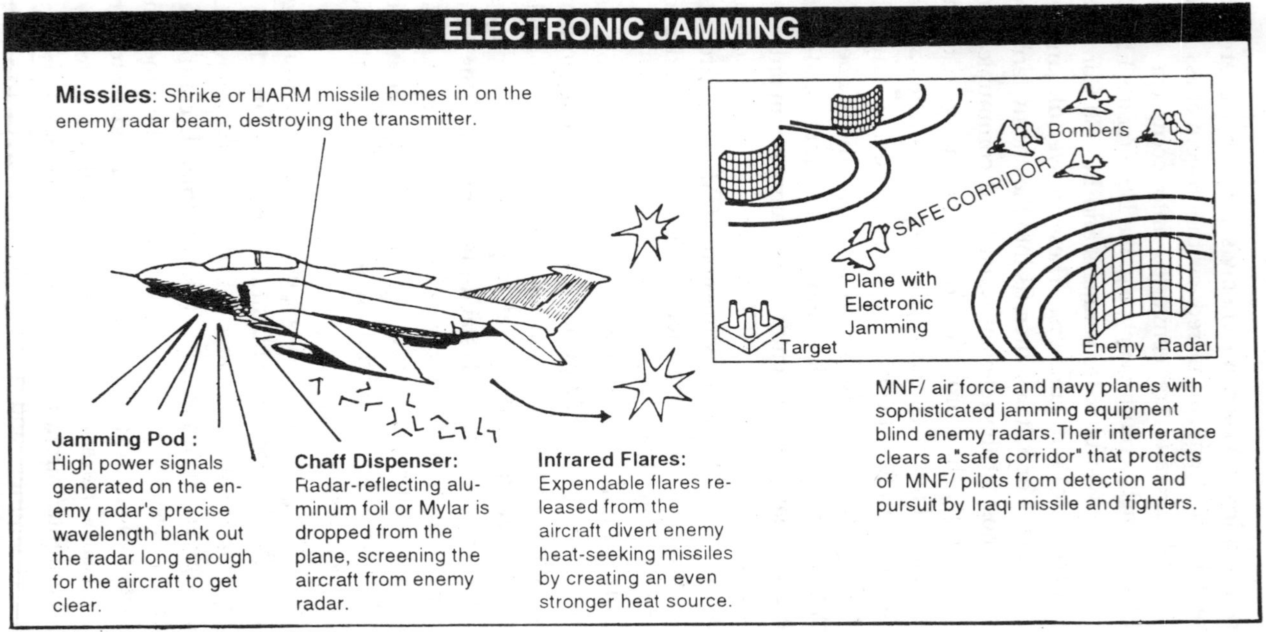

Figure 8
Electronic Jamming

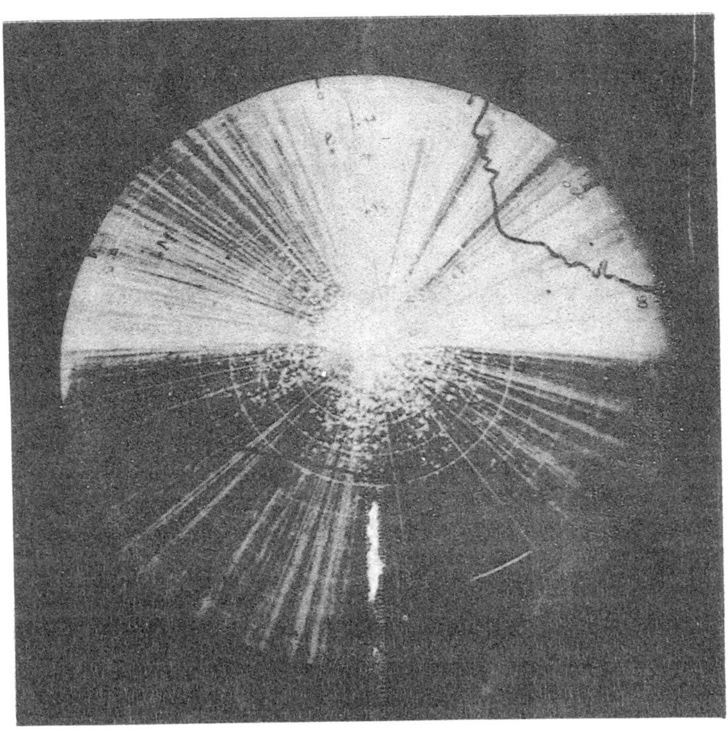

Figure 9
When Radar Gets blinded

missiles or radar controlled anti-aircraft guns but not for early warning or acquisition radars, provided these are jammed either by escort or stand-off jammer aircrafts.

If there are high places or hills available in a country from where one can have line-of-sight to the ground based radars in the enemy territory, the stand-off jamming can be carried out most cost-effectively and efficiently by suitably locating the jammers at such higher locations.The jammer aircrafts are very expensive systems. It is worth mentioning that the electronic systems of a modern fighter aircraft could constitute 30-60 % of its total cost which is of the order of 10 million to 30 million US dollars.

Electronic Chaff :

There is a very economical jamming technique using chaff which is made of millions of densely packed very thin aluminum foils of suitable lengths. A particular length of the foil is effective at a particular radio or e.m. frequency or frequencies around it. The length should be so chosen that it should be about half of the wavelength of the e.m. waves transmitted by the victim radar. If many radars are to be jammed simultaneously, various lengths can be judiciously mixed in each packet. Such a packet when dispensed from an aircraft, blooms immediately to form a cloud of these aluminum strips. The e.m. waves from the radar are reflected back by this cloud because these are made of a metal, like that of an aircraft. They appear on the radar scope as a powerful blip simulating the presence of a large aircraft. If large number of chaff packets are dispensed, they can create large number of blips amongst which it would be very difficult for the radar operator to identify the real aircrafts.

During 1943 in the WW II, when for the first time the German radar operators saw such a large number of blips on their radar screens, they immediately concluded that innumerable aircrafts have attacked. However, they also understood that such a large number of blips could not be due to the aircrafts and therefore, they concluded that, for sure, their radar was malfunctioning and thus they could not provide early warning. That day the allied forces using 791 bombers successfully demolished the Hamburg port under the 'umbrella' of this chaff and they lost merely 12 aircraft. If the Allies had not used the chaff, like in previous missions, as per past experience, they would have lost at least 36 aircraft. This resulted into the saving of 24 bomber aircrafts and pilots by dispensing only 2 tons of aluminum chaff. ECM is thus beneficial, and wonderfully cheap.

The use of chaff does not entail generation of electromagnetic energy as the chaff only reflects the energy radiated by the victim radars and hence it is known as 'passive electronic counter measure'. Therefore, it is a lighter and cheaper ECM system. Soon after the invention of chaff, an ECCM had to be found for radars to enable them to function despite the chaff. Such an ECCM device was soon invented and is known as 'moving target indicator' (MTI). Such a radar can find out whether its target is moving or not. If it is not moving it is not shown on the screen. Since a chaff cloud does not move significantly, it is not shown on the radar screen and is thus rendered ineffective. Chaff is getting out of date.

Role of Chaff Against Tracking Radars

A tracking radar of a SAM (or AAM) follows the target automatically by locking-on electronically to the reflected pulse, also called echo, from the target. The antenna of the tracking radar would then look towards the direction of the aircraft wherever it goes and the system would 'look' at the distance where it is moving. Thus the tracking radar always knows the direction and distance of its target and keeps on tracking it. The SAM system guides the missile only after getting locked- on to the target because then only the system knows the target that is to be attacked.

The 'lock' on the target aircraft can be unlocked by a series of chaff packet dispensed by the target aircraft. The unlocking takes place inspite of Moving Target Indicator (MTI) of the tracking radar of the SAM because the chaff strips, immediately after being released, remain in motion in the early seconds, therefore they look like a moving aircraft even to the MTI (Moving Target

Indicator) of the tracking radar. Since the chaff, due to large number of foils, produces a much stronger blip, the tracking radar gets deceived by their motion and strength, and shifts its lock from the aircraft on to the chaff. However, only in one or two seconds, the target aircraft, due to its speed, moves ahead whereas the speed of the chaff cloud becomes too slow. Then the MTI of the tracking radar is able to identify the chaff cloud as an stationary object and rejects it and gets unlocked from it, but it has to start its search for the moving aircraft all over again. The missile guidance system cannot order the launch of a missile unless it is locked-on to a target and can maintain the lock-on till the missile hits the target. Thus a chaff packet, costing only few dollars, renders a SAM system, costing few million dollars, ineffective. Thus cost- effectiveness ratio of an ECM device can be very high. But of course, as expected in the cat and mouse game, there are more modern tracking radars with ECCM devices that cannot be deceived by 'chaff', but can be jammed by active ECM devices.

Infra-Red (IR) Flare

Some SAMs such as SAM-7, SAM-9, SAM-13 etc., which were in the arsenal of Iraq, use heat seeking missiles, instead of radar guidance. Such missiles use the heat from the engine or exhaust of aircraft for homing on to it (aircraft). In the electromagnetic spectrum, the heat waves are neighbours of light waves but have lower frequency. The frequency of red colour waves are lower than those of other colours. Heat waves occupy a frequency spectrum which is just below that of the red waves, and hence are called infra-red (IR) waves.

The way chaff is used to deceive a radar, IR flares are used to deceive (IR flares are packets of solid fuels which can burn very intensely) the missiles which are guided by thermal sensors. When IR flares are released from the target aircraft, they get ignited soon after they come out of the aircraft and since the heat they produce is far more than that from the aircraft engine or exhaust, they attract the IR homing missile on to themselves and therefore away from the aircraft. Both the chaff packets and IR flares are generally dispensed by the same dispenser. Most of the modern fighter and other military aircrafts of advanced nations, which have to enter the enemy area, are fitted with these chaff/ IR dispensers. IR flares are small, light weight and very economical. Protection provided by such dispensers for penetrating a strong air defence system is not as safe as active electronic counter measures because various successful ECCM have been produced and are available. For IR homing missiles, the counter measure is either active IR jamming or IR flares. A combination of active ECM and chaff/IR flare dispensers provide sufficiently high safety to the aircrafts penetrating into very strongly defended air defence system at a reasonable expense.

Electronic Counter-Counter Measure (ECCM)

Both types of jamming described above viz. noise and deception are called electronic counter measures because they, with the help of electronic devices, disrupt the functioning of the enemy's electronic devices e.g. a radar. Counter-counter measure helps in maintaining effective functioning of the electronic systems despite being jammed i.e. despite the counter measures.

When deception jamming is being attempted, say on an enemy radar, the enemy might take suitable action to clear the deception and if he is technologically capable, he may succeed. For example, if he changes the pulse repetition frequency (prf) (not its e.m. frequency) of his radar, the artificial or deception blip may suddenly jump (no change in the blip from the real aircraft), thus giving itself away. Similarly, against the noise jamming, the enemy radar may use a special Dicky Fix Receiver or change its transmission e.m. frequency (not the prf) and reduce the noise effect. An example of SAM possessing such an ECCM known as 'frequency-agile radar' is Roland SAM system which Iraq possessed. The noise jamming effect would vanish from the radar scope till the jammer can find this new frequency and jam it again. Such an action by the victim to foil the counter measure action is called electronic counter-counter measure (ECCM). And if the jammer had a fast searching and tuning device or he knows the new frequency by other means, then he could again jam the radar which has changed its frequency to avoid jamming. Such an action by the jamming agency is known as ECCCM.

Moving Target Indicator - Immediately after the WW II, counter- counter measure for chaff was invented, based on the fact that chaff does not travel except at the speed of wind whereas aircraft does, and this ECCM which is based on 'Doppler effect' is known as moving target indicator (MTI) in a radar. The chaff is therefore not as effective against the modern early warning radars with MTI facility and acquisition radars, but is still effective against some tracking radars.

Sword-Shield Syndrome - If the jammer uses some technique which can make the counter-counter measures of the enemy ineffective, the process is called electronic counter-counter-counter measure (ECCCM). Indeed, it is like the development of swords and shields. If one can develop a sword, powerful enough to cut through any shield, the fencer with such a sword would easily win the duel. This is counter measure. To get over such a situation, the enemy would be forced to develop a better shield for his survival and possessor of the stronger shield would have greater chances of victory. This is counter-counter measure. Again, the one who develops even stronger sword, which can cut through the stronger shield, would snatch the victory. This is known as counter-counter-counter measure. And thus this never ending race of using stronger sword and even stronger shield goes on all the time.

Development work always goes on in the field of technology to produce more and more effective systems. As seen from the shield-sword example, a superior technology definitely enhances the chances of victory of the person who wields either a superior sword or a superior shield. It goes without saying that the person knows how to fence as well as, if not better than, his enemy.

Radar Warning Receiver

A modern radar warning receiver (RWR) warns the aircraft

pilot that an enemy's radar is either scanning or has locked-on to his aircraft and also gives direction of the threat. It can also sense frequency band of the e.m. waves radiated by the radar, some latest versions of RWRs can even evaluate the exact (e.m.) frequency of the transmission. It can also switch on the jamming transmitter (ASPJ) to jam the missile guidance radar automatically at the appropriate time. Thus an RWR is an extremely useful passive device i.e. it does not radiate any e.m. waves (except the insignificant natural heat radiations) and yet detects and indicates other radar's e.m. radiations and can also take necessary actions. The advanced versions of RWRs have a built-in library which stores the characteristics of important enemy radars so that it can also identify and indicate the specific type of radar which is either scanning or locked-on.

A modern RWR can be interfaced with a centrally controlled computer which would be actuated by the RWR when a missile is aimed at the aircraft. Suitable evasive action either to jam the missile guidance radar (if jammer available) and/or dodge the missile can be taken. The RWR, on its own, can also start dispensing chaff packets or IR flares (if a suitable chaft dispenser is carried) depending upon the type of missile chasing the aircraft. These protective ECM measures are automatically taken whereas the pilot, in the meantime, locates his target, fires a rocket, bomb or air-to-ground missile, destroys the target and returns to his base safely. All the while the escort or stand-off jammers continue to automatically jam the early warning radar, acquisition and missile guidance radar as well as the air-to-ground communication systems.

Elint Systems for Effective ECM

In peace time, every vigilant nation is busy in collecting relevant characteristics of the different radars and other electronic systems of its adversaries, actual or potential. These characteristics could be radio frequency of transmission, pulse repetition frequency, pulse width, antenna rotation rate, location, on-off timings and type of pulse modulation etc. The electronic order of battle, thus obtained, gives vital information about the enemy's capabilities and intentions; and tremendously helps in planning counter measures against the electronic systems, within a desired time frame, during a war. Such information is collected with the help of electronic intelligence (ELINT) systems, which are specially designed radio receivers.

Any sensible nation would change the frequencies and other important characteristics of its electronic systems, including their locations, during a war. This creates immense difficulties for the attacking forces because the information collected by their electronic intelligence system, during peace time, suddenly becomes of little use at the crucial time, but it is nonetheless useful because updation is easier and faster than ab initio collection. Hence collection of electronic intelligence is necessary even during war for taking realistic counter measures against the enemy's electronic systems. In such a case, unless the new frequencies etc. are found out, the enemy's electronic systems cannot be jammed during initial phases of the war. And one can not launch a war with desired safety, unless one has the information of enemy's electronic order of battle during war time. Air chief Marshal Neil Cameron, Chief of Staff of RAF had realised this problem and had stated that earlier it was considered essential to launch air attacks against enemy's tactical and strategic targets, but in modern wars, it would be essential to attack radars and SAM targets before other targets. But due to sudden change of electronic order of battle, it becomes a challenging and formidable task.

Under such circumstances, to avoid heavy losses initially, various clever tactics or devices are employed to gather such information. Before war has broken out, when the atmosphere is war-like and the tension mounts, mock attacks are carried out and the enemy, if he is not clever enough, may start using the new frequencies, considering that the war has begun. These frequencies are intercepted by ELINT systems, recorded and stored for use. In Kuwait war, the MNF had launched a few of such mock attacks on 15 and 16 January 91 i.e. before the commencement of the actual attack on 16 January 91, the ultimatum date being 15 Jan 91. As expected, Iraq changed it's

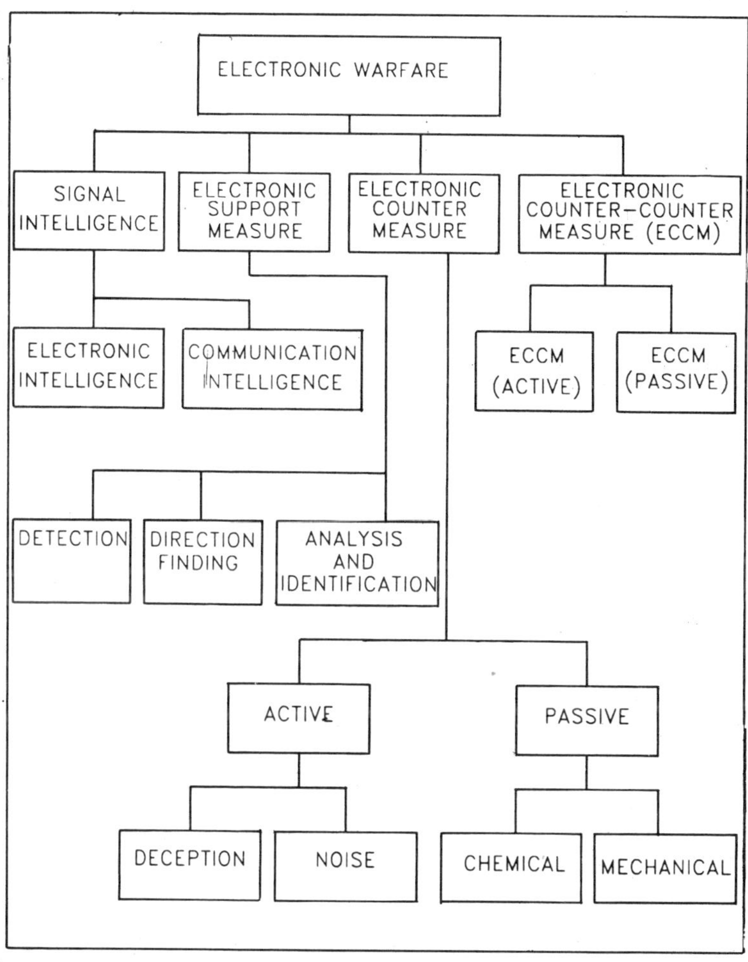

Figure 10
Family of Electronic Warfare

electronic order of battle to the wartime one and the MNF, with the help of ELINT/COMINT aircrafts and satellites, gathered all such information required for carrying out the ECM and ECCM effectively.

It is worthwhile noting that on one hand it had changed the e.m. frequencies of its radars and communication systems to wartime e.m. frequencies fearing a real attack, but on the other hand when, later on, the attack really took place, it took the attack as a harmless exercise ! Creation of confusion in the enemy's camp is one of the age old tactics of winning a war.

ECM & ECCM in Air Combat

During an attack mission if an adversary sends his interceptor aircrafts to foil the attack, the escort or stand- off jammers first jam the enemy's Ground Controlled Interception (GCI) radars and the associated early warning radars so that the fighter controllers sitting on the radar scope are not able to see the targets on the radar scope. Simultaneously the air-to- ground communication channels used by the GCI fighter controllers are also jammed, so that even if they can see some of the aircrafts occasionally, they cannot communicate with their interceptors. The fighter escort aircrafts accompanying the attack mission would also immediately engage the enemy interceptors, which somehow encounter them. If the number of enemy interceptors is equal to or less than the fighter escorts and the target is not far off, the attack aircraft would, most probably, continue the attack mission. If their (interceptors) number is more and the attack aircrafts have air-to-air missiles or guns, they may abandon the bombing mission and either run back or start fighting or combating to help their escort aircraft.

All the aircrafts which have active self protection jammers will also start jamming the air interception radars of the enemy interceptors. One who is able to first carry out effective jamming, chances of his winning the battle suddenly increase manifold. The successful jamming would make the victim forces virtually blind and deaf, subsequent to which the victory depends upon the efficiency of the pilot and capabilities of the aircraft and it's armaments. Thus it can be seen that ECM/ECCM cannot win a war by themselves, however successful they may be. But air power or military power can win a war, though not without the use of ECM/ECCM, in a modern war scenario.

Attack on Radars : An Extreme Counter Measure

If a nation has effective and sufficient number of jammers, every attack mission should be accompanied by escort jammers. A safer and surer, though more difficult, approach would be to first destroy the missile systems and radar/communication centres so that it will be easier to establish air superiority in the enemy's air space and number of jammers required to be carried can be reduced and the bomb load increased.

Let us first consider the attacks on radar units. One has to follow two approaches to do this. Firstly, since it is normally difficult to recognise and locate targets while flying at low levels (for safety), a direction finder is required which can give the direction of the prospective victim radar with high enough accuracy. An accurate direction finder functioning in the radar killer aircraft is a high-tech and expensive system. Secondly, one should have air-to-surface missiles with sufficient range which can ride on the victim radar's radiated beam and destroy the radar antenna/system completely. The missile should be so powerful that along with the antenna, the turning gear or the transmitting system also gets destroyed. Damaging the antenna alone would not serve much of the purpose as it can be repaired quickly.

If the target is protected by AAAs (radar controlled) and low level IR-SAMs and not radar guided missiles, the anti-radar missiles should have a minimum kill range of 4-6 km because these armaments have a kill range of about 2-4 km. Hence the mission should be so planned that the attacking aircrafts should launch their weapons from a distance of 4-5 km of the targets. If one does not have suitable rockets or air-to-ground missiles for IR homing SAMs, one would carry and dispense Chaff and IR flares and attack the AAAs (radar controlled) or IR-SAMs by entering the kill range of the IR-SAMs.

Guided Missiles

Laser Guidance - Precise aiming is done with the help of a laser, TV or IR guided missiles possessing a suitable kill range. A laser guided missile, when launched from a fixed wing aircraft, would attack the radar with high precision but needs another aircraft, though at a safer distance, to continuously illuminate the target radar with an appropriate laser beam, as the laser guided missile rides on the laser beam which is reflected from the target. On the very first night of the attack, two formations of Apache helicopters of USA had fired the 'Hellfire' laser guided missiles at important Iraqi air defence radars like Flat Face, Spoon Rest etc.

TV Guidance - in case of TV guided air-to-ground missiles an operator in another aircraft has to provide the guidance from a safe distance. Such a missile has a TV camera in its nose which transmits a movie picture of the target area (exactly like normal television transmission) to an aircraft at a stand-off distance, generally other than the launching aircraft which turns back to escape the air defence system. The control aircraft has a TV receiver which shows the moving picture of the target on its video screen to the controller or pilot. By looking at the TV picture, an operator (controller or pilot) in this aircraft controls the missile from a remote distance i.e. sitting in the aircraft it can change the direction of the missile so that it hits the target. TV guidance process can work only during day time. One of the versions of the American air- to-ground missile 'Maverick' works on TV guidance system.

IR Imaging Guidance - Night vision goggles have been developed with the help of which a pilot can see up to a distance of 10 to 12 km within a limited cone of 20° - 40°. Such a device can help the pilot in launching of a guided missile but only under certain conditions. For poor light conditions or foggy or smoky conditions, an IR imaging (IRI) camera is installed in the nose of the missile and the IR picture of the target area taken in complete darkness is transmitted to the control aircraft similar to that of a TV guidance system. With the video picture (IR picture is converted to video picture) of the target available, an operator can guide the missile precisely, exactly on the same lines as TV guidance. One of the versions of Maverick missiles works on this technology.

Anti-Radar Missile

When the target radar is protected by a SAM system with a kill range of about 15 to 20 km, which is normal with the radar guided SAM systems, the attack aircraft must have an anti-radar missile (ARM) whose kill range is about 25 km, otherwise the attack aircraft would be shot down by the SAM before it launches its ARM or any other weapon. HARM, high speed anti-radar missile, a US product and ALARM, air launched ARM, a British product, both have a kill range of about 25 km. These were freely used by the MNF to destroy Iraqi SAM radars. Iraq had the ARMAT, a French anti-radar missile but they never used it during the war because they had lost the air superiority.

Wild Weasel Aircraft

The US has a fleet of F-4G aircraft whose role is only to destroy the enemy radars and are known as 'Wild Weasel'. F-4G aircraft has a radar homing and warning receiver APR-38 which controls the firing of HARM missile through a computer. Its direction finding accuracy is about 1 degree (a technological feat !). Its computer has a library which contains all the important characteristics of the enemy radars in the area. The Wild Weasel pilot is left free to do his other onerous duties and merely observe the APR-38 which automatically executes all its functions. The APR-38 type of receivers provide immediate information regarding various electronic systems of the enemy and such tasks are called 'Electronic Support Measures' (ESM). Britain, for the purpose of this war, had modified its Tornado GR-1 aircraft for the 'Wild Weasel' role. The Tornado GR-1 has a radar homing and warning receiver and a computer with the help of which it can fire anti-radar missile ALARM from a distance of 25 km. There is one counter measure against the attack of an ARM viz. switch-off the victim radar for about a minute or so once the missile has been launched, and this would let the missile go astray. Although HARM has no counter to such a counter measure, ALARM opens a parachute and waits for the radar to be switched 'ON' and then attacks!

Stealth Aircraft F-117A

To destroy the enemy's radar itself is a highly effective ECM, albeit an extreme one. A fleet of Wild Weasel type of aircraft is expensive, but if some one can afford, they are very effective for both short-term and long-term ECM which would help strongly, economically and effectively in establishing air superiority over enemy's airspace. There is a supreme ECM against radar systems viz. stealth aircraft. To become invisible has been a very old dream of human beings, as per our Puranas, a repository of, inter alia, innumerable dreams, rakshasas could become invisible. Though radar is a very powerful eye, which can see upto hundreds of kilometers, modern technology has produced an aircraft which is invisible to radars, well almost. Such a stealth aircraft, F-117A, costs around 120 million dollars (as in 1991). No wonder, only USA can afford such a costly aircraft (the details would be discussed later).

ECM Through Tactical Flying

Now, let us discuss the ECM that can be achieved by planning the attack sorties tactically. Such a tactical planning makes use of two facts. The first is that the e.m. waves transmitted by the radar always travel in a straight line (well, almost). Consequently the radar waves emanating from the transmitter located on the earth's surface, move farther and farther from the surface, as they travel beyond a point, because the earth is curved (see the illustration). Thus if an aircraft flies low while approaching a radar, it is visible only when it reaches very close to the radar. The distance upto which a radar can see an aircraft is called its 'line-of-sight (LOS)' distance which depends upon the height of the aircraft and of the radar with no obstruction in between. Obviously the LOS distance increases with increase in the respective heights.

(a) Line-of-Sight
Radar Antenna
em Waves
Wave
A
D
S
Shadow
Region

(b) Line-of-Sight with Obstacles
Wave
50'
OBSTACLE
25'
24'
4 Km
10 Km
11 Km
27 Km
Wave
L.O.S.WITHOUT
OBSTACLE
50'

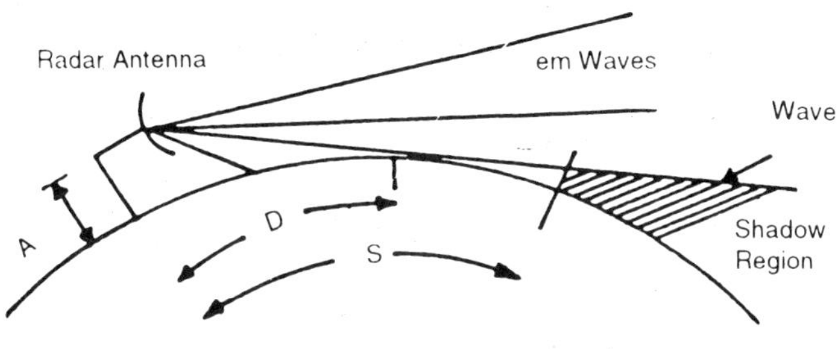

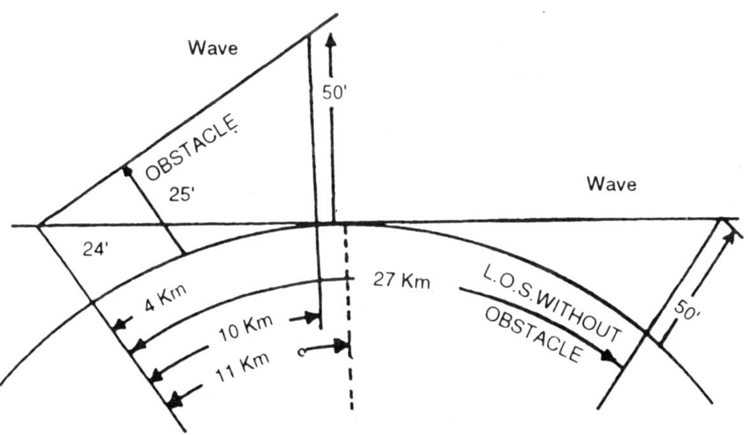

Figure 11

Radar antennas are invariably positioned at the highest possible point near a potential target in order to stretch the LOS distance and detect the sneaking aircraft from the maximum possible distance. The TV transmitting antennas are also located on high towers to increase their coverage. In the diagram (11a), the antenna height is 'A' feet. We assume that the rays emanate from the antenna dish in a straight line although they get slightly bent towards the earth as they encounter changing density of the atmosphere. Because of the height of the antenna, the rays reach all around upto a distance 'D' km, beyond which they start going away tangentially from the surface of the earth. Therefore, the closest one can fly clandestinely to this radar is 'D' km from it, and in fact slightly more than 'D' depending upon his altitude. This distance is the sum of two LOS distances, corresponding to the heights of the radar and the aircraft. For example, on a sea coast, if you stand at a point 50 feet above the sea level, the horizon that you could see would be about 16 km away. And if a ship was approaching you with its sail 24 feet high, you would be able to see the upper tip of the sail at a distance of 27 km (additional 11 km) Similarly, if a radar antenna is located at a height of 24 feet and an aircraft, flying at a dangerously low height of 50 feet, was approaching the radar, the aircraft would be seen by the radar from a distance of 27 km. A modern aircraft would take about 2 minutes to cover this distance. This time is not sufficient for older technology SAMs to reliably detect, track, lock and launch their missiles against it, whereas the aircraft could attack and return safely with a reasonable probability of survival. However, for the modern quick reaction capability SAMs, this period is more than sufficient and therefore the radars equipped with them cannot be attacked with impunity by merely flying low.

Flying in a Shadow

Unlike the foregoing example, most of the targets are not on the sea shore to get an unobstructed view. Generally on the surface of the earth, there are some high sand dunes, hillocks, clump of trees, buildings etc. and they cast shadows, thus reducing the maximum LOS distance and thereby helping the sneaking aircraft. Let us take the case of an obstruction 25 feet high, 4 km away from the radar. Computation determines that the new LOS distance would be 10 km which means that due to the shadow cast by this obstruction an aircraft flying at 50 feet would be able to come up to 10 km instead of 27 km earlier, from the radar, without being detected. An intruding attack aircraft can cover this distance in about 35-40 seconds. In such a short period, successful interception even by a quick reaction missile is doubtful. Pilots generally look for obstructions so that they can fly at a relatively safer altitude of 100 feet, although even from such heights it remains difficult to locate, aim, and attack the target successfully. The aircrafts which do not have any other electronic counter measures, are bound to take such low and hence dangerous sorties, in a desperate attempt to fulfill the attack mission.

To deter the aircrafts from sneaking at low levels, fast development of quick reaction low level missile systems, specially fitted with an IR guidance system has been undertaken since early eighties. Iraq has SA-7, SA-9, SA-13 and SA-14 IR homing missiles. To defend themselves against these IR guided SAMs and radar guided AAAs, the attack aircrafts carry IR flares, chaff packets and ECM systems. In such a warfare, if the counter measures succeed, the attacker would drop the bombs on the target and return safely (and if they fail, the attackers would become easy prey to the lethal SAMs). In any case, low flying has other deadly hazards as well; like ground fire from automatic rifles and even unguided anti-aircraft artillery (AAA). These weapons, being cheaper, are deployed in large numbers which compensate for their lack of accuracy and there is always a high probability of a hit against low level aircrafts. Even with effective ECM, to avoid a chance encounter with enemy's interceptor, some pilots (of the attacking aircrafts) prefer to fly low but not as low as they would have to fly without the ECM.

Flying at higher altitudes has its own dangers as well. For example, with the increase in aircraft height, it may become visible to more than one radar at the same time. Inspite of jamming of some radars, his position could be known through other radars, and interceptors from any suitable air base can be sent to intercept him i.e. it becomes vulnerable to additional forces of air defence system. The task of jammers and RWR increases and the probability of a mistaken identity also increases. Thus an attack from a higher altitude necessitates larger number of smarter ECM systems and fighter escorts. The US air force pilots believe in the high level attacks and hence their attack missions are suitably fortified with a large number of ECM and fighter escorts. Such high level attack missions need large resources but the probability of their success also increases because they can locate the target well in time, sufficient enough to adjust their flight path, aim at the targets and launch their weapons properly, without having to face the ground fire from the automatic rifles and AAA. Attack from a higher altitude may prove to be very dangerous if the enemy has powerful interceptor aircrafts, GCI system and medium range missiles with appropriate ECCM. But if the enemy's radars and communication systems could be successfully jammed and the enemy, in turn, is not able to successfully jam the attackers' airborne interception radars, air-to-ground missile guidance radar and IR imaging systems, the principle of high level attack is likely to result in a glorious victory with minimum casualties.

During the first 3-4 days of the Kuwait War, the British Tornado fleet with their time tested tactics, attacked the very dangerous targets from low levels and they lost one aircraft every day from the ground fire from rifles or AAAs. This happened despite the effective use of ECM because rifles and normal AAAs do not use radars for guidance and although have poor accuracy (there is no question of jamming these weapons by ECM), their numbers can compensate to some extent. The US forces attacked from higher levels, as per their established policy, well fortified with ECM and fighter escorts and in the same period they lost in total only one aircraft. Consequently, Britishers also changed their tactics in favour of high level attacks with very much improved rate of attrition.

From all the observations and studies, one principle comes out 'loud and clear'. The weapon of electronic warfare is extremely cost effective. An attack mission of aircrafts costing several million dollars can be made ineffective by successful jamming of its various electronic devices, air-to-ground missiles, precision guided munitions; and these jammers would cost a small fraction of the attack mission's cost. It is also true, vice versa, that an attack mission carrying appropriate jammers can render the air defence system, costing several million dollars, ineffective at a small fraction of the cost of the air defence system. Further, this advantage is not fixed but can be snatched by any party that has requisite technology and brain power. Here a comment deserves attention : Some military leaders consider expenses on EW systems as that on 'tail' and they claim that this reduces 'teeth-to-tail' ratio of the force. By now, it should be clear that EW systems belong to the 'teeth' and not 'tail'. If some act throws an oncoming missile away from its target, that act or system is surely a 'weapon', and if EW does the same, it is also a weapon.

It can be easily seen that effective and forceful attack with ECM, weapon system and suitable tactics are necessary for victory in a modern battle, whilst minimising the battle losses. ECM and ECCM act as effective force multipliers, par execellence. To achieve this, without undue risks, one needs to update knowledge about the enemy's 'electronic order of battle'. To understand the Kuwait War-91 better, we have to know the order of battles of both the sides.

> "In these days of frightfully expensive weapon systems, when a super power can easily dominate in any war, electronic warfare and information management provide a ray of hope for a nation with lesser resources but requisite high technology and brain-power to fight the injustice".
>
> - V.M.T.

"Technology itself may be to-day's primary air power theorist; that invention may, for the moment, be the mother of application."

David Mc Issac, Lt. Col. (Retd.).
Air University, USA

Chapter 5

High Technology Systems

HI-Tech of 1415 AD

In the year 1415 AD, the Britishers invented the longbow and in the battle of Agincourt defeated the French who were ruling supreme till then. The range of a longbow being longer than that of the conventional bow, they could hit the French soldiers with their arrows well before they would come within the range of French arrows. Longbow was a hi-tech weapon of early fifteenth century. In the stone age, invention of a stone-axe was probably the first hi-tech weapon which almost ensured the victory of the user of this weapon over his enemy who did not possess it, the enemy, by the way, included animals many times stronger than man.

Impact of technology can be drastic and far reaching for those who are not prepared for it. The battle of Agincourt is not a solitary example or in minority. The phenomena depicting the impact of technology on the outcome of fierce and bloody battles and wars can be traced from the age of stone missiles and stone axes of the most primitives of hominids till the age of nuclear bombs, Tomahawks and Stealth Aircrafts. The phenomena continues and the history of mankind is a witness to the fact that science-technology and warfare (also culture, material development, way of life) have always been intimately connected. Archimedes is one of the earliest modern scientist who had contributed to war efforts. Perhaps it can be stated with some exaggeration that foundations of modern physics were laid as a result of solving the ballistic problems by Galileo and Tartagha,the great Italian scientists of seventeenth century.

Inventions of the musket, flintlock and bayonet not only changed the course of wars, they also changed the composition and structure of the army. For example, during the reign of Charles VIII (France; late fifteenth century) the numerical strength of infantry was only twice that of cavalry showing the tremendous power of the latter. Only two hundred years later i.e. by 1700 AD, the comparative strength of infantry had increased to five times due to invention of musket etc. and partly due to developments in fortifications. Then came the canons which could destroy the forts with comparative ease and so on. Thus the relationship between technology and war has only grown thicker, complex and deeper all the time.

Donitz, the Commander-in-Chief of German submarine fleet during WW II, used a very bold and novel tactics by deploying the submarines, during nights, on the surface rather than submerged (for which they were designed), and dominated the Atlantic theatre during World War-II for a considerable period. However, only when the Britishers succeeded in using a 'magnetron' the high-tech ultra high frequency (in the electromagnetic spectrum) S-band device for radars (hitherto only much lower frequencies could be used for radars) and successfully detected the nocturnal surface movements of submarines, they could re-establish their supremacy over German naval fleets. This S-band radar was made possible by a dedicated team of scientists in Britain who developed the high-tech magnetron, well before the Germans could even dream of it. It is worth noting because Germans were the leaders in the field of radars and ELINT, as the first electronic surveillance flight was undertaken by Germans during 1939. Similarly many more examples of high-tech weapons leading to victory could be quoted from various wars, some have been quoted in this book.

In twentieth century, the Gulf War-91 has been a high-tech war, par excellence. In this war high-tech was deliberately given the most important role, unlike any other war in the past, because the US wanted to win the war in shortest period and to keep its war casualties the lowest possible. President George Bush, himself, had laid emphasis on this aspect because he did not want any opposition to the war from the anti-war lobby of US citizens.

Victory Through High-Tech

In 1921, General Douhet of Italy stated, "to win the next war, it should not be fought on the lines of the previous one, but with the help of high technology, it should be won with such a great speed that the enemy remains unable to respond." He further added, "if an aircraft is made the main weapon system, it can become the instrument of victory in the next war". It took 70 years to produce an aircraft, high-tech enough to become the main and effective weapon system in this war, and as a by product, prove his statement. It appears that General Schwarzkopf really grasped the philosophy of General Douhet and realised that the time has come for it to be employed.

Soon after Douhet, the famous British General and military thinker John Fuller had stated, "victory in a war depends 99% on high technology and 1% on all other factors". By other factors he meant offence, morale, strategy, tactics, courage, unity of aim etc. Perhaps John Fuller had taken recourse to hyperbole in order to make his point because the British Government was not giving sufficient importance to modernise the forces as per his suggestions. Whereas General Heinz Guderian had not only understood this principle but succeeded in modernizing the German forces, which gave them tremendous success during the initial stages of World War II. The USA has also believed in high technology, but had not been able to obtain an unquestionable proof of this principle till this war. It goes to the credit of hi-tech weapons that they inspired the US to deploy them even before they could provide a convincing demonstration over a test range.

In Gulf War-91, the MNF had used the latest hi-tech weapon systems including those which were not even tried earlier under field conditions. They were even criticized for such a drastic approach with a view that such green weapon systems would not be able to withstand the 'fire-test' of a war, and apart from their failure they would create additional problems.They were also criticized for using Kuwait war as a trial field for their green weapons. However, to-day hi-tech also inherently means high reliability in the production of systems also. It is kudos to modern hi-tech that the so called green weapon systems won the war confidently, that too, at such a hurricane speed that the enemy did not get enough time to respond.

What is a Hi-Tech Weapon System

With so many examples given, it should be easy to understand what is a hi-tech weapon. Continuous development goes on to improve weapons with the help of technology, as per the needs and ideas. Therefore a new and formidable weapon soon becomes outdated on development of a newer and stronger weapon. A high-tech weapon is one which is a successful product of the latest technology and has not been superseded yet. A high-tech weapon can be a hardware like F117A or Tomahawk or it can be an abstract system like information management system or electronic warfare tactics or techniques.

War Against Radars

By now, it is not surprising that advancements in the field of electronics and electronically guided weapons play the most important role in the wonders of high technology, consequently the modern ECM/ECCM systems play a dominant role in over-powering the enemy, specially because the electromagnetic waves are vulnerable to be interfered with from a distance. Looking at the important role played by air force during WW II, Chief of Air Staff of Royal Air Force, Neil Cameron had stated,

"Once it was important that air battles were fought before commencing any action on a battle front during a war. But it is my belief that now a battle against radars is most essential on which the success of air battle depends." Schwarzkopf made his plans, inter alia, on the basis of this principle and no wonder that Wild Weasels, electronically guided munitions, laser guided bombs and the epitome of the development in electronic counter measure (ECM) system - F-117A - stealth bomber aircraft,became the heroes of this Kuwait War.

Stealth Aircraft F-117A : The High-Tech Weapon System

The aircraft F-117A, has been in squadron service of USAF since 1983. F-117A along with many more modern weapon systems were used in the Kuwait War, yet they (F-117A) became the undisputed hero of this war. On 16 January 91, the MNF had 2430 aircrafts, out of which there were only 56 F-117A i.e. a mere 2.3 percent of the total. When on the first day of attack about 1500 sorties were launched, under the most hazardous conditions, against the most important targets, F-117A were allotted 31%, a lion's share of the total targets of the day. Why ?

We have seen that Iraqi air defence system was very strong. About 700 SAM systems and 10,000 anti-aircraft guns were guarding Iraqi and Kuwaiti air space with modern 'command, control and communication' systems for effective air defence. In such a well protected air space, penetration by conventional fighter/bomber aircrafts would have proved suicidal. Only those aircrafts that were equipped with modern Active Self Protection Jammers (ASPJ) and escorted by heavy ECM force could have penetrated the highly lethal air space but even then fear of high attrition was worrying the US. F-117A is a stealth aircraft which is practically invisible to a radar and therefore enemy can neither attack it with radar guided missiles or guns nor direct its interceptor aircrafts against it during nights. Therefore, F-117A could 'quietly' (stealthily) penetrate into their strong defence system, drop laser guided bombs on the designated targets with high precision and return to the base safely for another sortie. In fact, as the costs of defence

Figure 12

F-117A Stealth Fighter from the 416th Tactical Fighter Squadron, Tonopah Test Range Airfield, NEV., lands at the new desert shield home.

systems, specially those of fighter aircrafts, have been increasing prohibitively, a powerful lobby has been working against giving due importance to air power. In order to keep the investment in fighter aircraft low, a trend in developing countries has been to use older aircrafts with suitable upgradations. Another trend of keeping the platform itself simple and economical, and putting the developmental efforts in weapon systems, has been gaining strength. But the USA is rich and resourceful and where defence is a serious business, it went ahead in designing the most sophisticated and expensive platform, viz. F-117A, which became the real hero of Kuwait war. How does this extremely expensive F-117A become invisible to a radar ?

To detect the presence of aircrafts, a radar transmits a very narrow beam of e.m. pulse train. (See appendices for details). When the pulse train of radio waves transmitted by a radar strikes a metallic object like an aircraft, they get scattered in almost all directions. When a pulse train reflected by an aircraft reaches the radar receiver, after certain processing, the receiver shows the direction and distance of the aircraft on the scope (a screen like that of a TV) of the radar. If a target aircraft can absorb most of the incident e.m. pulse or is transparent to the e.m. beam (like wood or glass), it can become invisible to the radar. If the target aircraft can be made to reflect the radar beam in directions other than that of the transmitting radar itself, the radio waves would not reach the radar and then also the aircraft would become almost invisible to the radar. F-117A follows by and large the last principle.

Facetting - The scattering or reflection of radio waves from an aircraft depends upon its shape as well as on the materials that it is made of e.g. metals, composite materials etc. The fuselage and the wings are always rounded to generate maximum lift and to reduce the drag due to wind. Both these properties are essential for efficient flying. This rounding also results in reflection of the radio waves in many directions and some part of the radio waves are always reflected back towards the radar. This you can see for yourself that a round surface reflects rays, unlike a plane mirror, in many directions and also

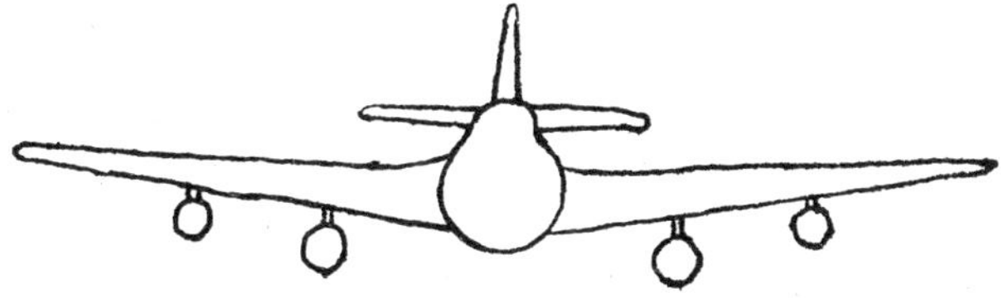

Normal Aircraft

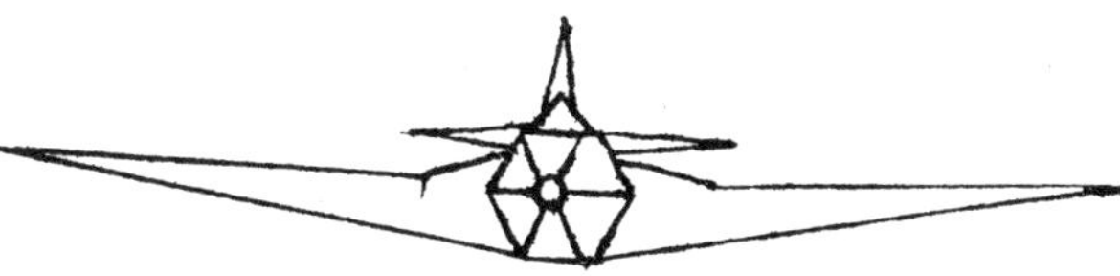

Stealth Aircraft

Normal Aircraft

Stealth Aircraft

Figure 13
Different Shapes of Stealth and Normal Aircraft

towards you, regardless of your direction with respect to the mirror (except in the blind area). If you can find a shining ball, the larger the better. you can see your image irrespective of the direction in which you are standing. In case of a plane mirror you can see yourself only from one direction i.e. when you are in a direction perpendicular to the mirror. Therefore, if the reflection towards the radar is to be reduced to the minimum, the rounding would have to be reduced drastically and plane surfaces would have to be increased. This process is called facetting. But facetting would reduce the lift produced by the wings and also increase the air-drag. This, in turn, would reduce the cruising speed of the aircraft or alternatively more powerful engines would have to be used. The cruising speed of F- 117A is less than other aircrafts of similar class despite more powerful engines.

Hollow but Solid - Almost all the surfaces of F-117A have been made plane and even the joints have minimal rounding. But then below the canopy, which is transparent to e.m. waves, there are many equipment which have zig-zag metallic surfaces and some of them might reflect the radio waves towards the radar. Hence the canopy in F-117A has been designed with special care. The canopy is made of 6 plane surfaces (instead of the standard oval) and then a suitable net or sieve of very thin metallic wires is placed below it or fine gold dust is mixed in the perplex material, of which the canopy surfaces are made of. The wire mesh is designed on the basis of special mathematical formulations so that the hollow mesh serves as solid surface to radar beams but not to the eyes. Now the radar waves would not enter through the canopy but get reflected in six directions, other than that of the radar, from the 6 plane surfaces of the canopy. Thus the canopy remains transparent to the pilot but acts as a combination of 6 reflecting plane surfaces to the radar waves.

When most of the aircraft surfaces are made plane, they ought to be perpendicular to some direction and any radio wave coming from that direction would be reflected back in the same direction. Acquisition radars or missile guidance radars are generally installed in the proximity of the target they have to defend. The attacking aircrafts, while approaching an important target, would be facing these radars. Therefore the plane surfaces in the stealth aircraft are so arranged that they are not perpendicular to the direction of the travel of the aircraft i.e. the front direction and also such that they generally reflect the radar beam towards the sky.

Just as the wire mesh placed inside the canopy is designed in such a way that its reflection should not be in the front direction, the facetting of the mouth of the engine (engine- intake) is done in the same way.

Radar Absorbing Material - Apart from the clever facetting, some aircraft surfaces are painted with a suitable layer of radar absorbing material, a material which instead of reflecting the radio waves, absorbs them (an example of passive (chemical) ECM). Thus the strength of radio beams reflected from F-117A towards the radar has been successfully reduced to about one thousandth of that from a non-stealth aircraft which makes them almost invisible. Any low level radar which could normally see a non-stealth aircraft, say from a maximum distance of 50 km, could see the F-117A only from a maximum distance of 8 km provided it would maintain the same sensitivity throughout. However it does not happen in practice. A radar designed for 50 km range, automatically reduces its receiver sensitivity for targets at shorter distances. Thus, practically it would not be able to see the stealth aircraft at all. However, as a good ECCM measure, a clever radar operator may manually increase the short range sensitivity of his radar, provided he has the prior knowledge of its arrival. A radar operator, for most of the time, can not maintain the same sensitivity for all ranges because then high power reflected by non-stealth aircrafts would create problems for the radar.

How Invisible is the Stealth Aircraft - Let us say, that a radar operator has increased the sensitivity appropriately when it sees a stealth aircraft at 8 km distance for the first time, he has to wait for another blip on the scope for confirmation because this blip could be due to some noise.. This takes about 3-5 seconds waiting time and in that duration the aircraft moves to a distance of 7 km. Rest of the distance to be covered by the aircraft is practically insignificant as the reaction time of a SAM system is of the order of 5 to 10 seconds, by which time the aircraft would have come to about 5 km and a radar's blind zone extends up to that distance below which the SAM cannot be launched. However while planning the attack mission, the F-117A could safely be kept at minimum distance of about 8 km from the radar. This minimum distance depends upon the distance between the target and the radar. If a similar computation is done for SA-3 ground-to-air missile guidance radar, it would be found that the radar, by maintaining the high sensitivity, could see a stealth aircraft only at about 3-4 km distance and at that time it can cause no harm to the aircraft as the aircraft would enter the blind zone of SAM III at the same time.

Increasing the receiver sensitivity has many associated serious problems which hamper the normal activity of the radar and its work increases manifold due to a large increase in the number of false warnings. It may even result into crashing of the computer or high increase in the probability of getting misleading information of targets. Finally, how would the radar operator know when to increase its sensitivity ? Indeed if the operator was to do this for all the time, he would see many targets that do not exist and would lose faith in his radar.

Radar Cross Section (RCS) - From the radar point of view, every aircraft has a property known as 'radar cross section' (RCS). The radar energy reflected by any aircraft towards the radar is an important design criterion for a radar and, inter alia, it is this RCS that determines the maximum range at which a given radar will be able to see that aircraft, an important performance criterion of any radar. The reflected energy from the aircraft is actually measured and, by using the laws of radio wave propagation, that area of the aircraft is mathematically evaluated which has reflected that much of energy. This effective area of the aircraft is called 'radar cross section'. This effective area is not directly related to the actual geometric area of the aircraft since many other factors influence the RCS. For a stealth aircraft the RCS is claimed to have been reduced to about one thousandth of its non-stealth version. The lesser is the RCS of an aircraft, the lesser is its visibility to a radar, and this reduction in visibility is related in an exponential manner to the RCS.

Infra-Red Stealth - Apart from a radar, the presence of an aircraft can also be detected by using temperature sensitive Infra-Red (IR) sensors, and radio receivers. IR sensors can sense a heat source like an engine or exhaust of an aircraft. Radio receivers can receive radio transmissions when made from an aircraft. It is possible to create 'electronic stealth' by stopping all radio transmissions (viz. communication and radar) originating from an aircraft. This makes the task of the pilot more difficult but the aircraft can still fly its missions. IR stealth is somewhat difficult. In the context of F-117A, two engines and their exhausts are really the two hot portions of the aircraft whose functions can not be stopped. Two heat shields have been put below the engines which help to reduce the temperature of the engine as viewed from below. Further the exhaust gases are made to travel along the internal surfaces of the aircraft before they reach the exhaust. Thus the hot exhaust gases heat up a very much larger surface area and thereby cause lesser rise in temperature due to quicker and larger radiation of heat. By reducing the heat density and hence the temperature rise, as viewed from downward direction, the strength of heat signature of the aircraft has been effectively reduced for F-117 A.

Electronic Stealth - If an aircraft radiates any electromagnetic wave, the enemy's radio receivers would be able to locate the aircraft. The pilot of this aircraft, therefore, has been told to maintain electronic radio silence i.e. he does not radiate any radio or radar waves in the enemy territory. By stopping radio/radar transmissions, difficulties would be experienced for navigation (e.g. doppler navigation system ought not to be installed on board), enemy aircraft detection (as done by a radar on board), checking weather en route (by a weather radar) and finding the height of the aircraft above the ground (by a radar or radio altimeter), because these are the functions that air-borne radars or radio sets perform. Instead of using a doppler radar system for navigation, F-117A uses a high performance gyroscope. This super gyroscope, though expensive, is highly accurate and does not use radio waves. As far as possible radio communication systems are not used within the enemy territory. The requirement of a radar altimeter is avoided by using a pressure altimeter, which is less accurate, but consequent risk is obviated by flying high which a stealth aircraft can do as there is no fear of being detected by enemy radars. Weapon aiming is done by IR or laser guidance systems, thereby avoiding the use of a radar for this purpose.

Low Probability of Intercept - For its attack missions, F-117A requires an accurate assessment of direction and distance of its target and that too during night time and under adverse weather conditions, which only a radar can do. But the e.m. waves, once emitted by the radar, would make their presence known to the enemy. To meet such an eventuality, a special stealth radar system was designed that can transmit radio waves which no enemy radio receiver may be able to receive. Such a wonderful technique is called 'low probability of intercept' technique which enables the radar to function with very low probability of detection by the enemy, one such method is called 'spread- spectrum'.

Spread Spectrum Technique - To enforce low probability of interception by an enemy and yet transmit radar signals, special modulation techniques are used. With such a technique, the radar transmissions would appear to enemy's intercepting receivers as noise only, that is also, if they can, somehow, receive them. When such a modulation is used, the total radar energy or power used for transmission is spread over a very much larger electromagnetic frequency band and hence is known as 'spread spectrum technique'. Due to spreading of energy in a larger frequency band, the power density of the radar energy gets reduced after such distribution, if the power transmitted remains the same. The wider the frequency band, the lesser would be the power density in the frequency spectrum, that means less power at each frequency in the band. For a given amount of butter, the larger the surface to be buttered, the thinner would be the layer of butter. In spread spectrum, the band width is increased to more than thousand times the conventional bandwidth and, therefore, the power density also gets reduced by the same order.

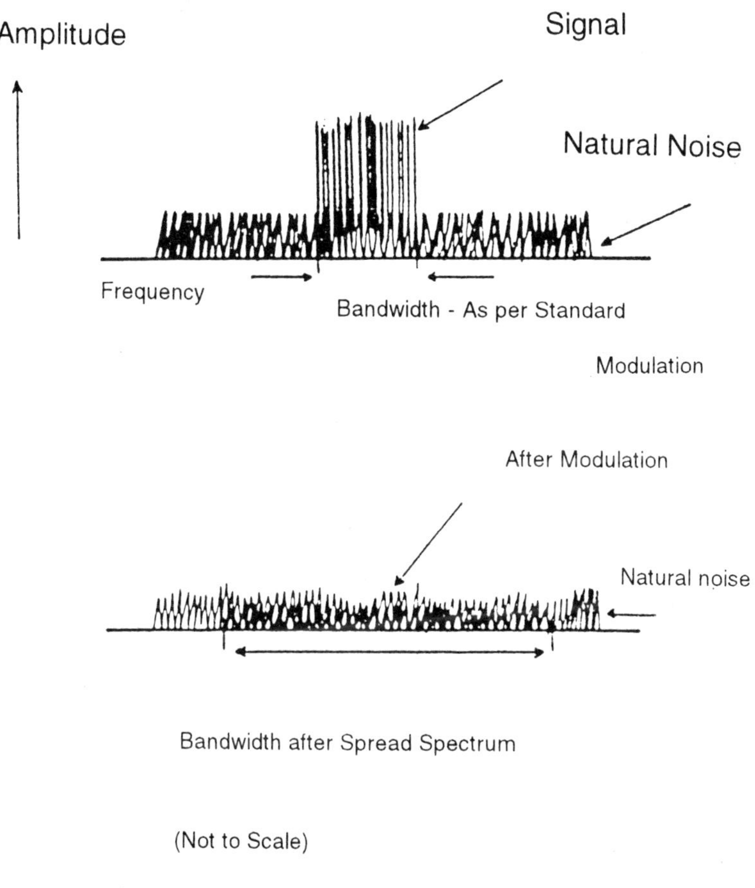

Figure 14
Spread Spectrum Technique

There is some natural e.m. noise present in the atmosphere at a particular density. The aim is to get the power density of the radar or communication signal after using spread spectrum far below the density level of natural electromagnetic noise generally present in the atmosphere. Therefore enemy's receivers cannot receive this radar signal because it would be below the ambient e.m. noise level, just as you would not be able to hear a whisper from one meter in a very loud discotheque. In order to receive it, the enemy's receiver, firstly, should know the e.m. frequency of transmission. In addition, the enemy's interceptor receiver should also know the exact modulation technique of spreading the spectrum, otherwise it would be impossible for him to receive such a signal.

This spread spectrum technique uses a modulation which when demodulated with anything other than the exact code originally used, appears as noise. This code is not only kept secret (though nothing can be kept secret for a long time with the present advancements, as the mathematicians would always be able to break the code, it is only a question of time), but also, it is essentially based upon generation of pseudo random numbers and then using them as modulating signal, the modulated signal thus produced by using the random bit sequence, behaves like noise. The enemy would therefore not even suspect the presence of the spread spectrum type of signal in the air, even after attempting to decode. Even then because electronic stealth is so important, use of such a transmission is kept to the bare minimum. Thus a radar using spread spectrum technique can locate a target and yet does not give itself away.

Keep the Enemy in Dark - In view of the above, F-117A is a stealth aircraft from all angles except to the sight of a man. This black coloured aircraft, therefore, generally attacks only during dark nights. Such a sophisticated and expensive stealth aircraft (costing $120 millions) is used only against militarily very important, heavily defended, and invaluable targets. The attack by a stealth aircraft has to be successful in the first pass (attempt) itself as the attack aircraft would not get an opportunity for the second pass because the aircraft would have announced his presence on the first attack with a bang. Even if the enemy is not able to pin point the attacking aircraft's location, its missiles and anti aircraft artillery (AAAs) would start firing blindly. Since the AAAs are large in numbers, there is a high probability that one of the shells might hit this aircraft. Therefore stealth aircrafts, like F-117A, must carry weapons which would get a bull's eye, every time, without fail. F-117A carries only two laser guided bombs, just weighing one ton each, hidden in its belly.

F117-A cannot carry more, as the extra bombs would have to be openly slung below the belly. This would necessarily increase the RCS of the aircraft hundred fold and severely reduce its stealth performance. Just to give you an idea, a normal ground attack aircraft, instead, carries 4 to 8 tons of ammunition. Indeed, what is important is not how much load of ammunition an aircraft can carry, but what can it drop on the targets with precision and at the same time maintain its own safety. Quality is more important than quantity, as usual.

Forward Looking IR System - All objects radiate infra-red waves and the frequency of such a radiation depends upon its temperature. Although at a particular instant, the ambient temperature may be regarded as constant in a localised area, there is always a difference in temperatures of every object, however small it may be. For example, in comparison to sand or clay, water gets warmer or cooler more gradually, metals get warm or cold faster. Without transmitting any e.m. waves like a radar does the IR sensors receive such IR radiations and are able to differentiate even minute differences in temperatures of different objects and give a picture of sufficiently high resolution which would be good enough for identification and aiming a weapon. Such a system in stealth aircrafts is specially useful for viewing a target in dark nights and is known as 'Forward Looking Infra-red' (FLIR) 'Weapon Aiming System'. This system is capable of projecting the picture (in artificial colours) of an area 20-30 sq. km in front of the aircraft on a TV screen. With the help of this picture, the pilot can lock-on his weapon to a selected point of the target, and launch its laser guided weapons.

A fixed wing attack aircraft carries out its attack from a low level and its speed is bound to be much faster than the bomb after it's release. Thus the aircraft, soon after the launch of its bomb, overflies the target, before the bomb hits the target. In such a situation, the attack aircraft has to reduce its speed because if the FLIR system starts looking at an area behind or below the aircraft, the pilot would not be able to look ahead, which can be dangerous at low levels. Therefore, an additional IR system known as Downward Looking Infra-Red (DLIR) Weapon Aiming system is installed in F-117A which automatically takes over the role of FLIR as soon as the aircraft crosses over the target. The pilot continues to look at the scene in front and the DLIR system handles the task of aiming the weapon automatically. DLIR has a laser designator also to point a laser beam on the target as directed by the pilot. The laser beam is switched on only a few seconds before the bomb hits the target so that even if there is a laser warning receiver near the target, it would not get sufficient time to react. The laser guided bomb hits the target with a fantastic accuracy (within 1 metre of the aim) and destroys the target. If the target is large in size or tough, both the bombs are launched in a quick succession, or a special size bomb can be developed and F117-A be modified for its launch.

Normally, in an attack mission two F-117A go as a team to ensure perfect accuracy in aiming and assured destruction of the target and safe return to their bases. The second aircraft would launch its bombs only if the first aircraft has somehow failed to get a bull's eye or the target has not been fully destroyed. It is highly remarkable that even after making thousands of dangerous sorties in this war, not even a single F-117A got lost or damaged.

Four Dimensional Navigation : At Right Time in Right Place - For navigating inside enemy territory, with minimum risk of exposure, pilots use 'Ground Based Mission Controller' to decide the safest route. This can be a time consuming and a complex task as the pilot would not like to fly very close to radars that can be avoided and yet he would like to spend minimum time in the enemy territory. For a stealth aircraft which is penetrating into unknown enemy territory in dark nights, an accurate and reliable automatic navigation system is absolutely essential. This is achieved by a gyro-based navigation system which is so accurate that the pilot can reach the target within a few seconds of the estimated time. Its navigation system is four dimensional i.e. it navigates the aircraft in latitude, longitude, height and time. For launching the attack, the pilot is only required to survey the area, some seconds before he reaches the target.

F-117A pilot is required to look at the IR picture only about half a minute before the time on the specified target by using FLIR in the wide angle mode. Having identified the target, he switches the FLIR to the narrow mode for better identification and locking the IR camera on to the target. Appearance of the target within a marked circle on the FLIR scope means that if the bomb is launched at that moment the target would be hit accurately. The bomb is first guided by the FLIR and then by the DLIR system till the bomb hits the target, always a bull's eye. During Kuwait War, a laser guided bomb destroyed a huge semi-underground structure by entering through its window.

Catch the Enemy Sleeping - F-117A does not need escort jammers in all of its missions, nor does it need escort fighters in most of its missions. As it reaches a target stealthily, its attack is always a surprise to the enemy. Surprise has a very big advantage in battles and is a major contributor to victory. Stealth as an ECM system is really supreme not only because it is passive but it makes the aircraft invisible to the enemy radars and therefore it prevents the enemy from using any radar or I.R. guided weapon against the aircraft.

Being only 2.3% of the total air force, the F-117As were given 31% of the total targets chosen for the first 24 hours of this war and by the time the war was over the figure rose to 40% of the targets. Out of the total 110,000 sorties carried out by all the MNF aircrafts, F-117As carried out mere 1600 flights. This means, with very much less number of flights (about 1.5% of the total flights), more targets (about 40%) were destroyed, with no

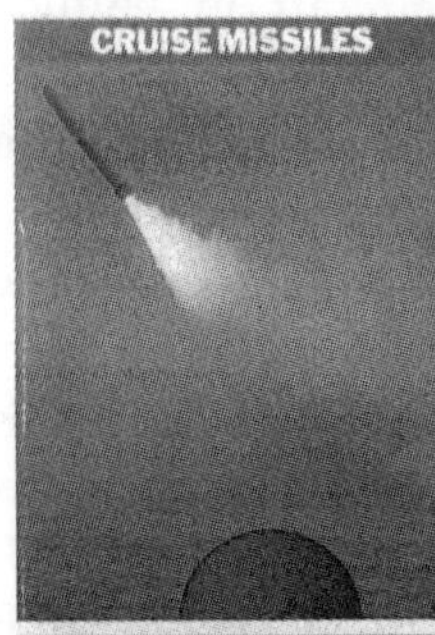

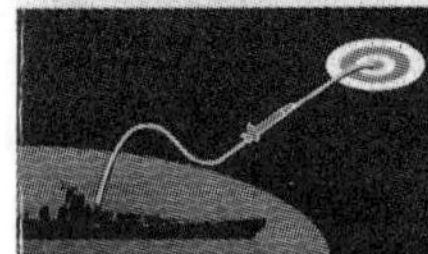

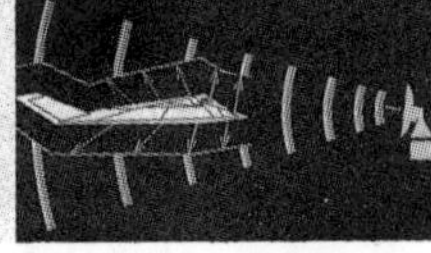

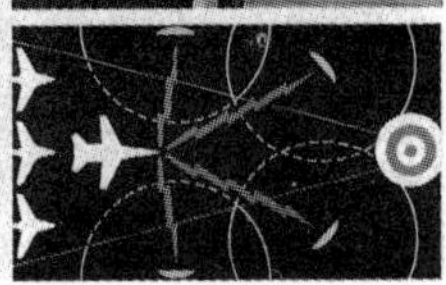

PURPOSE:
Long-range attack
USES:
Launched from ships and submarines
DISTINCTION:
Can fly under radar; 1,500-mile range
COST:
$1 million

Under a moonless sky over the Persian Gulf, 100 of these missiles initially blasted off from U.S. warships on a 700-mile flight to Iraq. Their TERCOM radar system compared landmarks with prerecorded maps to guide them to their targets. They struck nuclear, chemical and biological facilities.

PURPOSE:
Long-range precision bombing
USES:
To penetrate air defenses undetected
DISTINCTION:
Extremely low radar profile
COST:
$106 million

Taking off from bases in Saudi Arabia, 27 of these single-seat twin-engine planes were the first aircraft to hit such targets as command-and-control centers and fixed Scud missiles. The plane's radar-evading Stealth technology, based on shape and materials, proved highly successful in the gulf.

PURPOSE:
Confuse or disable enemy radar
USES:
Carried by the Navy's EA-6B Prowler, the Air Force's F-4G Wild Weasel, EF-111A Raven and EC-130H Compass Call
DISTINCTION:
U.S. has the most advanced systems now deployed
COST:
For a Prowler, $32 million

The latest electronic-countermeasure systems have been placed on new planes as well as some of the oldest in the U.S. inventory. In the gulf war, ECM aircraft were among the first over Iraq and Kuwait, jamming air-defense radars and crimping their ability to detect intruding planes.

Figure 15

loss of aircraft and no abortive mission, a performance, unparalleled so far of ground attack capability for any aircraft.

The First Night : Ends With A Bang And Not With A Whimper

Iraqi air defence could have proved most dangerous during the first 24 hours. Only if the MNF air attack missions during that period could hit Iraqi targets (mainly the 'command, control and communication' organ of the war machinery , high/medium/low level radars, missile guidance radar etc.,) then onwards Iraqi air defence system would become less effective. Hence the attacks during the night of 16-17 January 91 were most important and most risky too.

In a way, it was a "catch-22" situation. Attacks were extremely risky, and the risk could not be reduced unless attacks were carried out. Only hi-tech weapons could resolve this dilemma safely. Three types of high-tech armaments were used during that crucial night. First, laser guided bombs were dropped by stealth aircraft F-117A; the second, Hellfire laser guided air-to-surface missiles which were launched by the attack helicopter to hit the low level early warning radars, and the third, Tomahawk, long distance surface-to-surface cruise missiles were fired from the Naval ships. This use of hi-tech weapons demonstrates that the Iraqi air defence was extremely dangerous.

Only Dumb Miss, PGM Gives A Deadly Kiss

Laser guided bomb is a precision guided munition (PGM) because it can be guided to hit a target with a very precise aim. A PGM can be a 'glide' munitions or it can also have a motor in it. Unguided bombs, now-a-days called dumb bomb, hit their targets with comparatively large margin of error, and consequently cause unnecessary destruction by falling at wrong places. During the Vietnam war, when hundreds of dumb bombs failed, a few laser guided bombs were enough to destroy the Than Hoa bridge. Therefore, a laser guided bomb is more cost effective than a dumb bomb. Cost of a GBU laser guided bomb used by F117-A is about 30 thousand dollars. When compared with a dumb bomb's cost of 2000 dollars, GBU laser guided bomb is definitely cost effective. Thus, apart from tremendous increase in bombing efficiency, one can see another crucial advantage gained through the use of PGM, viz. tremendous reduction in destruction of unaimed objects and population.

The most tragic aspect of a war is killing of innocent civilians and destruction of their property. Any action that ameliorates this situation or reduces such killing and destruction is a humane act. If we cannot stop a war, let us at the least stop this inhuman killing of innocent civilians. To achieve such a precise guidance, the target is illuminated by a laser beam and the bomb rides on the laser beam reflected by the target and hits it with an accuracy of one metre. Illumination of the target by the laser beam is carried out from another aircraft or from the ground based designator (sometimes persons from Kuwaiti resistance movement helped in such tasks). The laser designator can also be in the same aircraft (as in the case of Apache helicopter or F-117A). Immediately after releasing the bomb, the attack aircraft will overfly the target and then the laser transmitter should have the manoeuvreability to look downward and then backwards so that it continues illuminating the target till the bomb actually hits it. A British bomber aircraft, Buccanier, also has such capability to launch laser guided bombs and designate the target for its own use as well as for other aircrafts. Buccanier aircraft has a TV camera, a laser range finder and a designator located in its pod. Some modern aircrafts like A-6E, F-111 and F-117A use an IR camera and laser designator, both mounted on the same platform and suitably harmonised so that the laser bombs could be launched even during nights. Tornado aircraft has a 'thermal imaging and laser designator' which does the same task.

Electro-Optically Guided Bomb

In an electro-optically guided bomb (EOGB), there is a TV camera and a communication system which transmits a picture of the target area to the attack aircraft at stand-off distance. An operator looks at the video picture and by sending appropriate instructions through the communication channel, guides the bomb to the target, when at close range, for high accuracy, the bomb homes on to the target by using mm (milimetre) wave terminal guidance radar in the active mode. As a TV camera or a low light level TV camera can take a picture only during day light or low light level conditions; for night attacks, instead of TV cameras, IR cameras are used by mounting them in the missile itself. In electro-optical guidance, after dropping the bomb it is not essential for the pilot to keep flying towards the target, i.e. it can turn in any direction once a lock on TV or IR camera has been achieved, only the aircraft has to fly high enough so that the target remains within his line-of-sight (LOS). Only the American F-111 and F-15E and the British Tornado GR-1 could launch such electro-optically guided bombs.

Laser Guided Bombs with More Punch

F-117A had mainly launched GBU-10, Pave Way II laser guided bombs (LGB) to destroy a number of very important targets. GLU- 109B was specially developed for destruction of reinforced underground bunkers. Its casing is made of hardened steel and the nose has an armour piercing shape. This bomb can penetrate 9 metres of reinforced cement concrete structure and then explode to disintegrate it.

Effectively, statistically speaking, one laser guided bomb is equivalent to about 40 dumb bombs but costs, only 10 to 15 times the cost of a dumb bomb. During the war when they found that even the special 1000 kg LGBs were not able to destroy some highly fortified bunkers, USA got another special 2136 kg bomb developed on top-priority basis, during the war itself. On the penultimate day of the war such bombs were dropped by two F-117A on a very important bunker. It was probably a bad luck that one bomb hit a bunker meant for civilians. This was perhaps due to wrong intelligence/misunderstanding or a clever step taken by Iraq of changing the use of bunkers between military and civilian personnel. This certainly was not a deliberate act because this is one of the extremely few cases where civilians

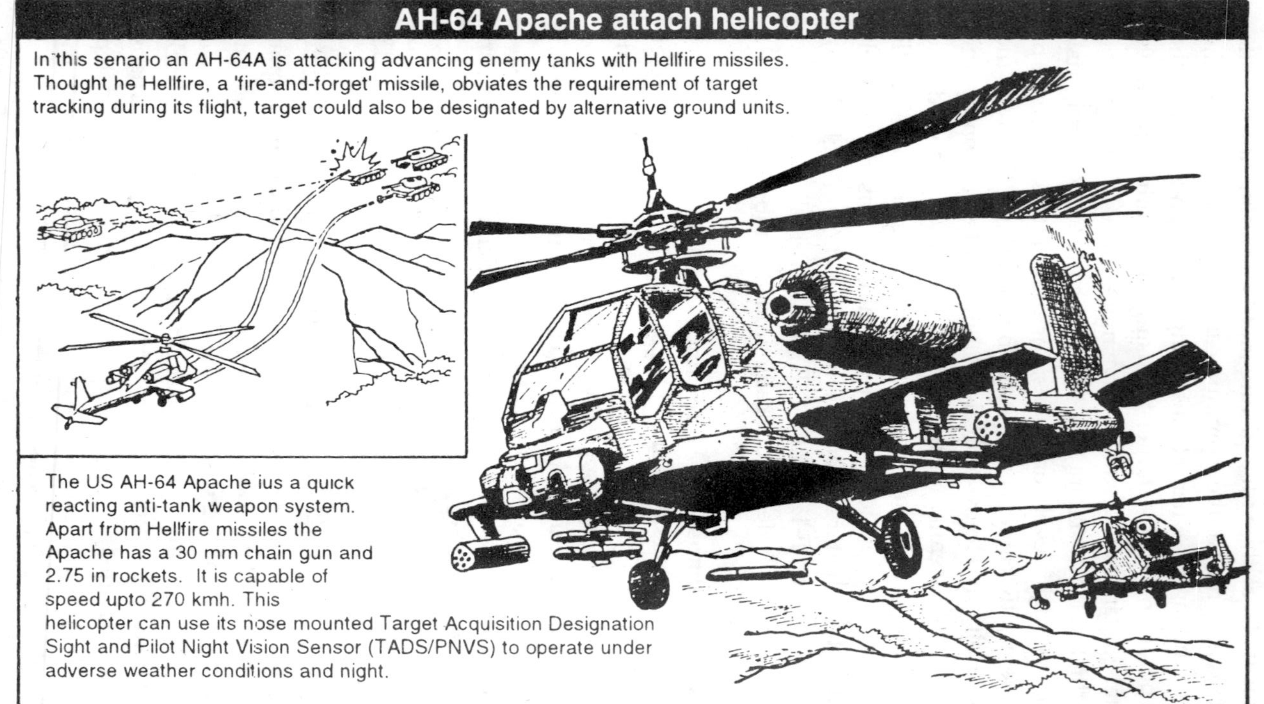

Figure 16
AH - 64 Apache attack helicopter

w...e hit by the MN... without serving any useful purpose. It is worth noting that the low ratio and numbers of civilians killed by MNF in this war is a record. It was part of a strategic plan in the interest of the U.S., not to have civilians as war casualty.

French Pilots, from their Jaguar ground attack aircrafts, during the Desert Storm, launched about 60 AS-30, air-to-surface laser guided missiles, with the help of a very modern ATLIS pod, against the Iraq's Air Force stations, bridges, important bunkers, ships etc. ATLIS pod automatically tracks a target after acquisition and designates it with a laser beam.

Hellfire Missile Burns the Enemy in Fire of Hell

Hellfire AGM-114 is a laser guided air-to-ground missile which is launched mainly by Apache, Super Cobra and Black Hawk helicopters. This can hit and destroy targets like communication centres, radars, bunkers, armoured vehicles, reinforced concrete buildings, ships, tanks, helicopters and slow moving aircrafts during day or night, provided the target is laser illuminated by its own aircraft or an escort aircraft. One advantage of this missile is that it can automatically lock-on to the target after being launched. Consequently it can be safely launched from a helicopter, hidden behind a hill up to 300 m height or a building, a clump of trees etc. After being launched, the missile automatically climbs to a height of about 300 m, searches for the target being illuminated with a laser beam by an escort aircraft from a safe distance, locks on to it, and then dives on to it to attack.

Helicopters having laser designator, like Apache, can illuminate the target themselves and launch Hellfire. The advantage of their slow speed is that , like for fixed wing aircraft, they do not need a special illuminator which can look both downward and forwards. Helicopters like Apache can carry up to 16 Hellfire missiles, at a time, in each sortie. Since it is not advisable to spend long time in the enemy territory, every pilot wants to fire quickly all the 16 Hellfires on 16 targets and run back. But this missile does not permit the self designating helicopters to fire and forget because the targets have to be continuously kept illuminated till the missile hits them.

Generally, it would take long time to fire a large number of missiles from one aircraft, because one would have to wait till the previously launched missile hits it's target, and only then can second target be illuminated and second missile launched. The missiles would get thoroughly confused if different targets are simultaneously illuminated and a number of missiles are launched in a quick succession. In Hellfire system, there can be different types of laser beams for different targets and each missile can then be made to home-on to a particular target illuminated with a particular type of beam. Different beams are created by using different modulations on the radiated beam. The laser receiver in the missile can differentiate between these modulations and respond only to the pre-set modulation. In this mode, the missiles, pre-set with different modulations on each of them, are launched at an interval of about 8-10 seconds. After the first target is destroyed, the pilot changes the modulation of his designator to suit the second missile and so on for other targets. Thus a helicopter like Apache takes only 2-3 minutes (otherwise 4-6 minutes) to launch 16 missiles on say 16 tanks and destroy them. The modulation of a laser beam is changed by changing its pulse repetition frequency as the laser receiver of Hellfire has the capability of receiving only one type of modulation and each missile is designed to match a specific modulation corresponding to the specific target designator. Eight seconds are sufficient enough to change the modulation of a laser designator, change the aim and provide guidance.

A Hellfire missile costs about 53,000 Dollars and a modern tank may cost anything from 5 to 10 million Dollars. Further an Apache aircraft may cost about 10 million dollars. Unless it can surprise the enemy and catch him unaware, it is essential to provide safety to this attack helicopter in the enemy airspace i.e. to have air superiority. This helicopter would, otherwise, be an easy target for enemy fighter aircrafts.

The range of Hellfire missile is about 8 km. It means that most of the low level IR guided SAMs, AAA units and other targets, protected by such air defence weapons, could be considered as good targets for Hellfire missiles. In other words, as long as the range of enemy's surface-to-air weapon is less than 8 km, and there is air superiority, Hellfire can safely by used in the 'stand-off' mode, against such targets.

The weight of a Hellfire missile is only 45 kg and that of its warhead is 10 kg. Due to its special construction and very high speed, it can destroy the strongest tanks. It can penetrate 3 m into land and thereafter 70 cm of reinforced cement concrete structure. Hellfire and laser guided bombs, both, with the help of very superior guidance systems can get bull's eye and that too during the darkness of night, provided the aircrafts have suitable night vision devices.

Tomahawk Scores A Goal At 1100 Kilometers

The third weapon used on the first night of attack was Tomahawk, a surface-to-surface cruise missile. It goes only up to a maximum altitude of about 30 m over sea surface or land and approaches its target at a speed of 885 kmph, hence it is called cruise missile. The reasons of its flying low are the same as those for the attack aircrafts in enemy territory - not exposing itself to the enemy's radars at the same time giving a surprise to the enemy. This cruise missile could also be termed as a stealth weapon with the consequent advantage of surprise attack. Its RCS is about one hundredth of an aircraft, due to which it reduces its chance of being seen by a radar many fold when compared to any conventional aircraft. It thus reaches its target with a bang without giving any clues to the enemy and catches them unawares.

Tomahawk, essentially a weapon of the US Navy, has a range of about 1100 km. The US Navy could safely launch it from the ships positioned at the far end of Persian Gulf, Red sea and eastern Mediterranean sea on any target in Iraq. Even after travelling such a long distance, this missile could hit a target with an accuracy of 6-10 m, with a warhead of 1000 lb. or multiple smaller warheads.

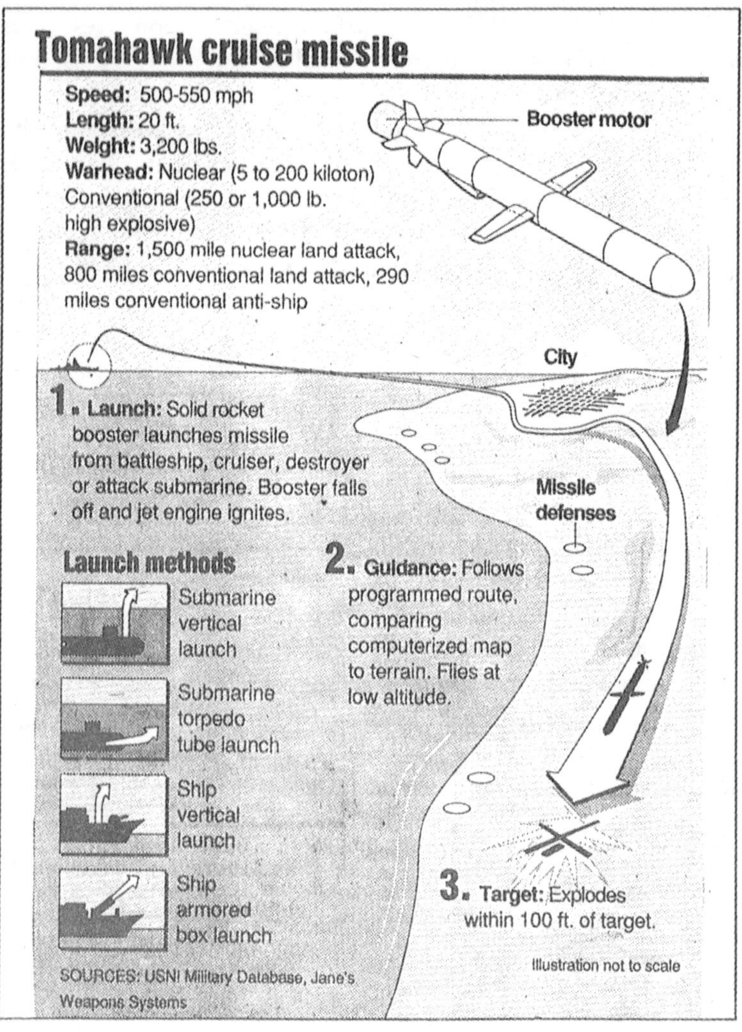

Figure 16
Tomahawk cruise missile

During its travel over sea, it is guided by inertial gyroscope (similar to F117-A with similar advantages) and then during its travel over the earth's surface it navigates by measuring the terrain contours with the help of a height finding radar (unlike F117-A), before reaching the first check point. The estimation and checking is done with the help of a computer which contains the path contours in its memory. It also corrects the flight path of the missile at every test point after identifying them. This process is called 'terrain contour matching'. Since a very high level of accuracy and also higher level of stealth is desired in the last phase of its travel, the task of terminal guidance is performed by a special computer by 'digital scene matching area correlation' technique, without using a radar. The digitised map of the target area is stored in the computer's memory and an appropriate sensor (a TV camera or IR camera) is used to take a picture of the area, once the missile arrives there. Both the pictures are then correlated by the Tomahawk's computer to know its own position within an accuracy of 5-10 m, on the basis of which it corrects its flight path and hits the target with a bang. This war has thus provided some glimpses of a war in future, including push button warfare. It also proved that computers can fight a war, but loss is always of humans.

Stealth Missile : Expensive, Safety, Surprise and Accuracy

The Tomahawk missile cannot be easily intercepted by enemy's air defence system since it flies at low altitudes and its Radar Cross Section (RCS) is negligible due to its small size and small diameter (70 cm). The safety also increases as its flight path is intelligently chosen by directing it through the weaker zones of the enemy radar coverage or taking the advantage of shadow zones formed by sand dunes, small hills or ridges. The element of surprise reduces, if more than one Tomahawks are guided to traverse the same path, as the enemy, by keeping a watch in that direction , can shoot them down by their low level quick reaction SAMs. During early phases of the war, due to such a mistake, Americans lost one or two Tomahawks to Iraqi missiles. Once the mistake was realised and rectified, the success rate of Tomahawk in smashing the targets was 95%.

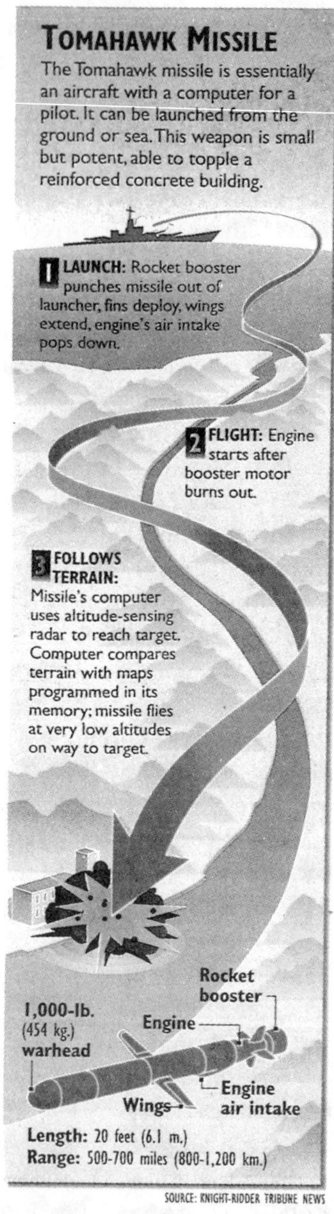

Figure 17
Tomahawk Missile

The high accuracy and safe travel was obtained through high- tech guidance, but at a very high cost. A Tomahawk costs about 1.2 million Dollars and its warhead weighs only 450 kg (equivalent to 1000 pound bomb). Each Tomahawk requires a digitised map for each target which takes a long time to prepare, and hence aiming it on a mobile target, like a mobile Scud launcher, is rather difficult if not impossible even for a Tomahawk.

Grand SLAM

On the very second day of the war, another hi-tech weapon, SLAM, was launched on an important target. SLAM (Stand-off Land Attack Missile) is an air-to-surface, 'fire and forget' missile. Two A-6 ground attack aircraft took-off from the John F Kennedy aircraft carrier in Red Sea and launched a SLAM each from 80-90 km from the target and 'ran back' immediately. Two A-7 guide aircraft, guided these missiles to the targets, from a safer distance.

For guidance in its early phase of travel, a SLAM uses a satellite based navigation system called 'Global Positioning System' (GPS) (described later) which provides a positional accuracy of about 10 m. SLAM has an IR imaging camera as well as a TV transmitter. In the later phase, about 60 seconds before hitting the target, the IR camera located in the missile takes IR movie of the area and passes it to the guide aircraft through the TV transmitter. The operator in the guide aircraft looks at the IR movie of the target area in his TV and guides the missile by sending the instructions on the same communication link and achieves precision aiming. Since the launching aircraft can fire the SLAM from a long distance of 90 km while flying low, it is not visible to the enemy radars i.e. it would not only remain out of range of the enemy missiles but also out of the range of the acquisition and guidance radars excepting that of early warning radars. The guide aircraft has to be in line-of-sight with the target, but can be at a safe distance. SLAM is an excellent example of high- tech weapon system with high precision, long range and night capability. Therefore for its deployment in this

Figure 18
Bunker Blaster—A tomahawk cruise missile, armed with a
conventional warhead.

war, its development was accelerated. It costs about one million Dollars.

Maverick Missile : Not Maverick At All

During Kuwait War, Maverick missile was found to be so effective that it was being launched at a rate of about 100 per day. It costs about 100,000 Dollars, double the cost of a Hellfire but with a warhead of 135 kg, thirteen times that of Hellfire. The US aircrafts F-16, F-15E and A-10 were used to launch it and each of them could carry 6 of these missiles in each sortie.

Maverick AGM-65XG and AGM-65D, both are IIR (Imaging Infra Red) missiles. It has TV and laser guided versions also but they were scantily used due to their limitation of day time deployment. Before a Maverick missile is fired, its IR camera shows the picture of the target area to the pilot on a TV screen. The pilot first looks at it in wide angle view (about 20 degrees), then in narrow angle view (about 5 degree),he locks the missile on to the target, launches it from a distance of about 24 km from the target and runs away. This missile also is truly 'fire and forget' type, as after being launched it automatically homes on to the target which was locked by the pilot before the launch, that too with a very high precision.

AGM-65G Maverick missile weighs 300 kg and has a larger warhead (135 kg) and that is why it was specially used against fortified bunkers. When it hits a target, mostly its speed is supersonic and therefore its effectiveness is also superb. AGM-65D weighs 225 kg with a warhead of 57 kg. It has an armour piercing shape and is therefore more suitable against tanks. The range of Maverick or other IIR missiles is basically limited by the sensitivity of its IR camera. Britishers had developed a sensitive Thermal Imaging And Laser Designator (TIALD) pod which further increases the capabilities of precision guidance missiles. Tornado GR-1, while carrying a TIALD pod, can become an electronic stealth aircraft. The US system LANTIRN (Low Altitude Navigation and Targeting Infra Red by Night) is also a similar system.

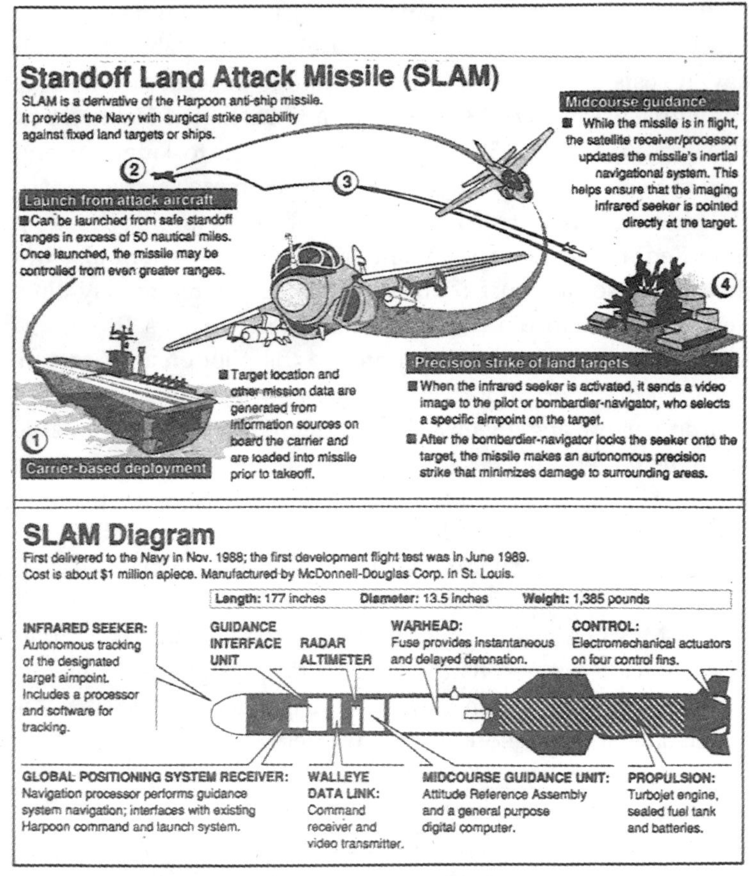

Figure 19
Acknowledgement : Vayu

Have Nap : Enemy's Nightmare

The air-to-surface precision guided missile (PGM), Have Nap, revitalized a 40 year old bomber, close to its death bed, and gave it so much fire power that it did not have even during its youth. Have Nap, AGM-12A has a navigation and guidance systems similar to that of AGM-84 SLAM excepting that in this case the IR imagery is sent to the aircraft which launches the missile, not to another guiding aircraft. In addition the terminal guidance is done by the missile itself. The pilot locks the missile on to the target with the help of IR picture. Have Nap has more powerful warhead (320 kg) than that in SLAM and its range is 110 km. The old B-52 G bomber can carry eight such missiles and in comparison to other aircrafts it dropped largest bomb load during this war. Though this missile gave such an infallible fire power to the old B-52G, which got its strength to perform this extraordinary task due to establishment of air superiority, without which its sorties in the Iraqi skies could have been suicidal as this aircraft cannot escape from any modern interceptor.

Patriot Saves Innocent Civilians

The most talked about missile, all over the world, was Patriot which intercepted and downed most of the Iraqi Scuds, Al Husseins and Al Abbasses. Although initially designed as a surface-to-air missile to intercept attacking aircrafts, quickly it was modified to serve as an anti-missile- missile (AMM) also. The political dreams of Saddam Hussein that he had nurtured on the strength of Scuds were, to some extent, torpedoed by Patriots. On April 25, 1991, Raytheon Vice President Skelly, in an official statement, said, "We believe that in Saudi Arabia, just under 90 % of Scud missile engagements resulted in destruction of the Scud's warhead". This was proved to be a wrong statement. Iraq had launched 38 Scuds against Israel and 43 against Saudi Arabia. None of them could hit the desired target, 11 of them fell down, on their own, without causing any harm, about 40 were intercepted by Patriots (this figure was found to be unreliable on later scrutiny), and only some fragments and

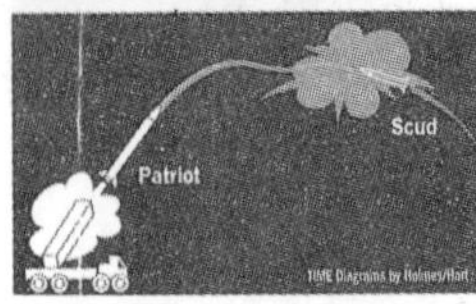

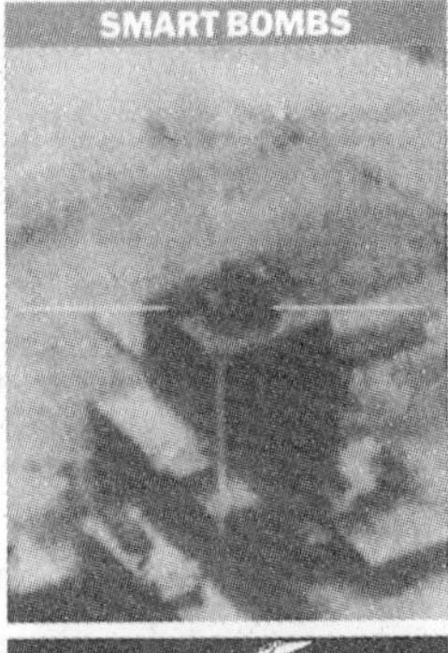

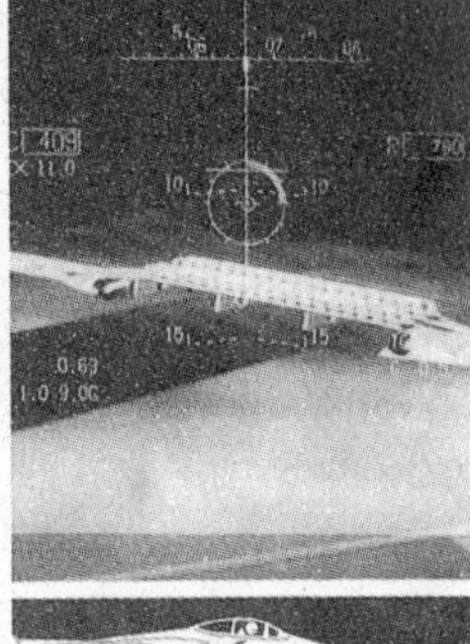

PURPOSE:
Intercept aircraft and missiles
USES:
Protection of ground facilities
DISTINCTION:
Has remote launchers and high accuracy missiles
COST:
$123 million

This system won high marks when a U.S. Army Patriot destroyed an Iraqi Scud missile in Saudi Arabia. A Patriot battery has eight launchers with four missiles each. Israel received two batteries in late December, but they were not yet operational. The U.S dispatched more, including crews, at week's end.

PURPOSE:
Precision bombing
USES:
Carried by most new fighter-bombers as well as B-52s
DISTINCTION:
Permits pilots to release bombs at safe distance from air defenses
COST:
Varies widely by type

The success of last week's air strikes was largely owing to the use of "smart bombs." Deployed in many forms, they are guided either by lasers, infrared or TV cameras. In one such system, a crewman can follow images relayed from the bomb and keep it on course toward its target by moving a joystick.

PURPOSE:
Light amplification
USES:
For aircraft, tanks and infantry
DISTINCTION:
Can amplify starlight 25,000 times
COST:
$200,000 for the fighter-bomber version

Night-vision goggles worn by fighter-bomber pilots, including those flying the F-15E Eagle that was used extensively in the gulf strikes, make objects visible at up to seven miles, even on dark nights. The device permits pilots to attack at low altitudes without using radar, which an enemy can detect.

Figure 20
Patriot System

splinters of 7 Scuds could reach the desired targets. Maximum damage was done by one such fragment, ironically caused due to interception by Patriot, which fell on a military barrack in Dahran and killed 26 soldiers. However, when in April 1992[1], Admiral David Jeremiah, Vice Chairman of the Joint Chiefs of Staff was asked by a reporter, " whether the Persian Gulf countries interested in buying a Patriot missile had been made aware of the fact that they do not function as successfully as it was reported in the begining ?" The Admiral replied, "Caveat emptor" (meaning buyer beware). There were many causes for failures of Patriot and for misreporting of a high success rate. One of the reasons was psychological warfare, to build confidence amongst the civilians (both Israeli & Arabs) and to let Saddam believe that his Scuds are not going unchallenged. Patriot was not really designed for a high speed missile like Al Hussein. An intercept of a missile means hit on the warhead, but while initial reporting, a hit any where near the missile was taken as an intercept.

Some weaknesses of Patriots experienced during the war have been made public. The first was that a Patriot hits its target missile but not its warhead, which may still cause serious damage while falling in fragments, as happened over Dahran. Secondly, some times (it happened quite often) due to false-alarms produced by its receiver in its self-guidance mode, the system launches its missile on non- existing targets. Thirdly, it intercepts a victim missile only when it comes very close to it. Fourthly, the miss distance at which Patriots were set to explode, were too high to be effective. Inspite of all this, 'Patriot' has demonstrated and proved the neccessity of an anti-missile missile and thus has given a glimpse of future wars.

J-STARS Shines for the MNF

The aircraft E-8A— J-STARS (Joint Surveillance Target Attack Radar System), meant for joint (army, navy and air force) surveillance, is one of the latest products of high-tech war machines. A J-STARS essentially consists of a radar (moving

---

1. Arms Control Today, November, 1992.

target indicator, MTI) to survey a vast area for locating moving targets, a synthetic aperture radar to locate static targets precisely, a computer system for processing the data related to various targets and a jam resistant communication system to convey the information related to the targets and to control the attack aircrafts. The larger the aperture of a radar antenna, the more sensitive it becomes and the more precisely it can see. Increasing the aperture or size of a radar antenna creates some difficulties, specially for an airborne system, therefore once again the magic of electronic circuity has come to the rescue in the form of 'synthetic aperture'. This aircraft does not merely identify and display the mobile as well as static targets detected by its radars but also provides necessary direction to own attack-aircrafts for destroying the important targets. The J STARS also proved to be highly effective in determining the extent of damage caused on targets as well as troops' movement, a task that AWACS cannot perform. It is surprising that J-STARS did not succeed very much in locating mobile Scud launchers, although it played a valuable role in detection of Scud launchers during the war. Its importance in ground battle is the same, may be more than that of AWACS in the air defence. The two J STARS flew 54 sorties totalling 600 hours and their contribution in the Operation Desert Storm was considered to be highly significant in achieving the lightning victory in the Kuwait War.

Wild Weasels Cut at the Roots

If F-117As laid the foundations for the air superiority, the Wild Weasels, the radar hunter and killers, formed the tangible structures on which air superiority was firmly established and then later on promoted to air supremacy. Prior to this war, the role of Wild Weasels i.e. to destroy the enemy's radar units, was played by a pair of aircrafts. One aircraft was used to locate the target precisely, whereas the other was to escort the first aircraft and drop a bomb on the victim target with a perfect aim. Serving the purpose of two aircrafts by one is another proof of the latest high-tech developments. If the kill range of HARM and ALARM would have been less than 24 km, it would have been difficult to safely launch the attacks and successfully destroy SA-3 type of surface-to-air missile units, perhaps ARMs were designed with such targets in view.

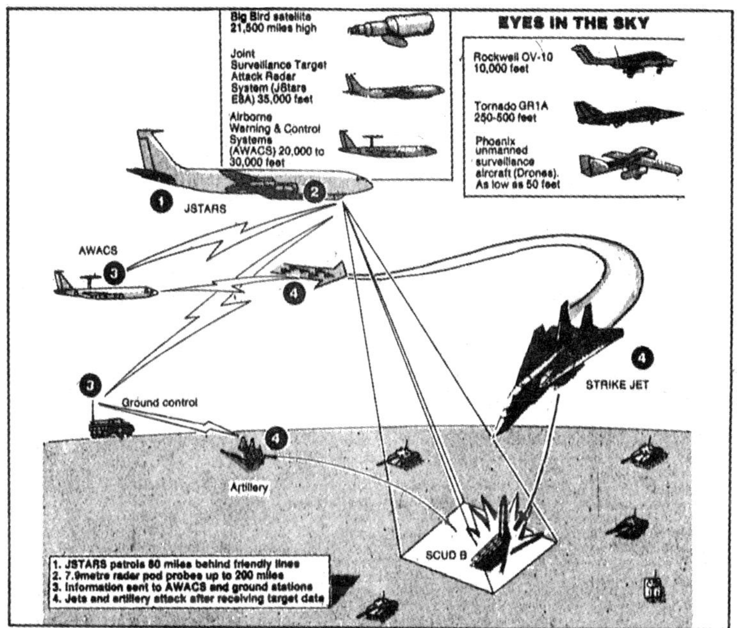

The all-seeing system

Figure 21

JP-233

Airport and their runways are difficult to be damaged by air-to-ground attack in a threatful environment as the runways are long and narrow with deep foundations and are made of special strength cement concrete. To make the task of attackers really risky, these important targets are well defended by AAAs, SAMs and interceptor aircrafts. How the strong electronic walls of defence are penetrated with jamming has already been discussed to some extent and would be further discussed shortly. How ground attack aircrafts destroyed strong underground bunkers has also been discussed. To destroy or severely damage such tough and well defended runways, the attack aircraft must first penetrate through the air defences, reach the target runway and get the target on the bombsight and destroy it with an appropriate weapon. The MNF had three weapons to perform such a difficult task - Durandal Bomb, JP-233 and MW-1. A standard laser guided bomb would not be effective enough.

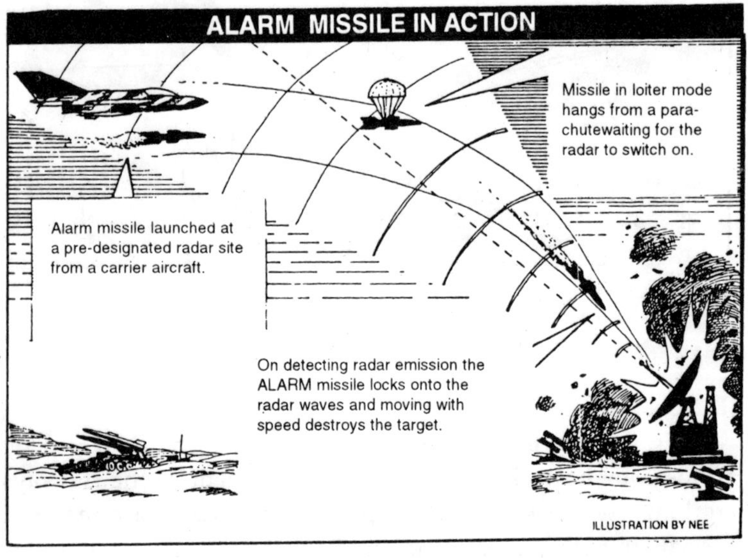

Figure 22

Durandal is a powerful bomb capable of digging holes on strong runways. However, for attacking very long and strong Iraqi runways, it was not considered safe in the early stages of the war, because a pilot, if he remains under heavy pressure, is unable to get a precise aim. MW-1 was used only by 10 Italian Tornado GR-1 aircraft. Therefore the brunt of attacking the Iraqi runways fell on British aircraft Tornado GR-1 and its bomb JP-233. JP-233 is a podded system and two such pods are slung below the belly of a Tornado GR-1. An attacking aircraft, for shelling the runways, is required to fly at a dangerously low level over the runway to drop JP-233. Thus British Tornadoes had undertaken a formidable task of very risky attack missions and that too from the very first night of the war. When launched, JP-233 pod releases 30 of SG-357 surface penetrating and cratering ammunitions on the runway. Thus a successful mission by four Tornadoes may cause up to 100 craters on the runway and leave a rather long and difficult task for the enemy to repair it. The repairs take a very long time and even after repairs such a heavily damaged runway would not remain sufficiently safe for the aircrafts. Though a difficult task, the Britishers had succeeded in damaging a large number of airfields, within 3-4 days of the war, with the help of this new high-tech weapon JP-233. This also helped the MNF in establishing their air supremacy.

Sparrow : In Reality A Hawk

The US AIM-7, Sparrow, is an air-to-air, medium range, radar guided missile with a maximum kill range of 40 km. It has all weather, day and night and all aspect, all heights capability. It's air-interception radar receiver works on 'inverse monopulse seeker' principle which gives it a very high jam resistance capability. The guidance for the missile is semi-active type in which the radar transmitter is in the aircraft and the receiver is in the missile. The radar waves reflected from the victim aircraft are received by the missile for guidance. Its digital signal processing technique also provides it some jamming immunity. Its effectiveness can be judged from the fact that, in the air combat, out of all Iraqi aircrafts downed by the MNF, 60% were hit by Sparrows. Fixed wing aircrafts of Iraq were shot by F-15C air superiority aircraft which have a very effective ECM system and use Sparrow missiles. F-15C had also shot down 5 MIG- 29 aircraft whereas not a single MNF fighter aircraft could be hit by Iraqi Air Force. It is known that capability-wise MIG-29 is superior to F-15C, then how did this happen?

Jam to Win : Drown in Din

Essentially there are two possible reasons for this surprising victory of F-15C over MIG-29. One is the successful jamming of Iraqi air defence radars (including destruction by Wild Weasels) viz. early warning, acquisition, missile guidance radars of SAMs and air-borne interception radars of interceptor aircrafts like MIG-29, thereby making the Iraqi air defence system blind. On the contrary, despite some attempts Iraq could not jam AWACS and other systems of the MNF. MIG-29 became blind for long ranges and hence could not use their AA-7 medium range missiles. Iraqi communication channels were also jammed to make them deaf and dumb in addition to blind. The air combat thus turned out to be an attack by a strong and capable force on a deaf, dumb and blind force. The first credit of establishing the air superiority should, therefore, go to 'jamming' rather than to F-15 and others who really shot sitting ducks.

Heavy Attacks Keep the Enemy off-Track

Heavy attacks have played a dominant role in the victory of the MNF. The advantage of heavy attacks is that damage inflicted on the enemy is assured and so heavy, that it is difficult, if not impossible, for him to recover in a short period. Let us say, if an air force station has been attacked heavily which causes serious damage to its power house, communication centre, air traffic control, runway, water supply system and some depots like ammunition, food, clothing and spare parts etc., it is impossible to make it fully operational quickly. Now if the same targets were attacked one at a time, the enemy could get them quickly repaired, one by one, and the airfield would not remain in-operational for long periods.

Heavy attacks need heavy resources, and carry larger risks, and therefore their planned, coordinated and safe implementation becomes an absolute necessity. The MNF planned for 2000 - 3000 sorties per day, on an average. The attacks were massive rather than just heavy. These flights were originating from Bahrain, Saudi Arabia, Turkey, Diego Garcia as well as from aircraft carriers positioned in Iran Gulf, Red Sea and Mediterranean Sea. There were different kinds of aircrafts for all kinds of functions-surveillance, reconnaissance, AWACS, J—STARS, refuelling tankers, air traffic controllers, laser designators, path finders, ECM escorts, Wild Weasels, air superiority aircrafts, attack aircrafts, bomber aircrafts and fighter escort aircrafts etc. All these aircrafts belonged to seven countries. These sorties were to be undertaken during nights also in all types of weather conditions prevailing in the desert region and that too after penetrating through the strong air defence system of Iraq. The planning, co-ordination and execution of such flying tasks was a serious challenge to any capable nation.

Blue on Blue, If no Clue

This is really surprising, pleasantly, that despite 2000- 3000 sorties a day in an extremely complex and dynamic environment, there was no 'blue on blue' i.e. not even a single collision or shooting down of a friendly aircraft. It indicates that the 'command, control and communication' for flight controls, planning of sorties and their minute to minute execution was near perfect. The 'Daily Task Order' was well chalked out and issued from Riyadh based Operations Control Centre (nick named Black-hole) to all the concerned units in time. What was the secret of such a phenomenal and unprecedented success ?

Knowlede is Power

As explained, it was a herculean, if not impossible, task to manage. Indeed such huge, dense and powerful attacks were unthinkable but for the high-tech information management that was under development since last few years. As the modern high-tech war systems are becoming more and more powerful and capable, their management is also becoming complex and needs hi-tech handling. It may or may not be true that the nation that would control the information, would win a war or would rule the world, but it is certainly true that without appropriate information management, now onwards, no nation can hope to win a war or rule the world. In a book like this one, only a brief introduction of the new hi-tech information management techniques can be given.

Information Management

Components of information management systems are :
1.   Sensor systems - to gather information.
2.   Communication systems - to allow the information to be shared.
3.   Computaion systems - to sift, analyse, categorise, organise, store and synthesize to make it useful in decision making.
4.   Presentation/interaction systems - to enable the users to use the 'decision support system' thus produced.

Sensor Systems

The sensor systems can be spaceborne, airborne or deployed on ground. A brief description for some of these (US owned) follows :-

DSP Satellites: Defence Support Programme Early Warning Satellite system consists of three satellites in Geosynchronous orbit over South America, the central Pacific and Indian ocean. The system can detect the trailing plumes of Submarine Launched Ballistic Missiles (SLBMs) or Intercontinental Ballistic Missiles (ICBMs) with the help of rotating IR detection scanners. Data from these satellites is relayed through defence communication satellite systems to ground stations at Buckley Field, Colorado and Alice Springs, Australia. The satellite system was used to detect and relay launch of Scud missiles in the Gulf War.

Defence Met Support Programme (DMSP) Satellites: DMSP Satellites gather meteorological data for making accurate weather forecast in support of military operations. The information can be beamed down to receivers on land, ship or aircraft. The data can also be stored or burst-communicated to stations, when satellite passes over them.

AWACS: Airborne Early Warning and Control System is an airborne radar system capable of depicting air defence situation within 500 Km of the operating area. The aircraft is a Boeing-707 modified to accomodate a rotodome on top, which houses the radar antenna. The air defence situation information is communicable to ground stations as well as fighter/support aircraft. A similar system is the E2-C Hawk Eye Early Warning System. The major advantage of these systems is their ability to pick up even the low flying targets (like strike aircrafts and cruise missiles) and feed the information into air defence information network in real-time.

J-STARS: This system has already been discussed earlier. The massacre of Iraqis, despite being in well entrenched fortifications, by well directed artillery and precision guided munitions from aircraft, is a testimony to the efficacy of this system.

Communication & Computation Systems

MILSTAR: Military Strategic/Tactical and Relay system is a constellation of eight satellites which provides secure information medium for worldwide communication. The MILSTAR can be used by ground stations, aircrafts and ships. This is by far the most impressive communication network available anywhere in the world.

DSCS III : Defence Satellite Communications System satellite is the current system to provide worldwide information exchange for the US and its NATO allies.

NAVSTAR GPS : NAVSTAR Global Positioning System is a satellite system for providing navigation information to mobile forces. It allows an unlimited number of military users to passively receive precise position, velocity and time information anywhere on or above the surface of the earth. The NAVSTAR constellation consists of 28 satellites, 10 out of which are back-up (standby). The accuracies provided by this system are remarkable; 16 metres in position, 10 metres/sec in velocity, and 100 nano second for time. This system has been used with a telling effect in the Gulf War by the combat planes of the US.

JTIDS : Joint Tactical Information Distribution System is a digital communication system for transmitting data and voice communication. It is available for Army, Navy as well as Air Force operations for air/ground communication.

Decision Support Systems

The Decision Support Systems (DSS) are the information systems which directly assist the man in the field (be it a commander or a soldier) to take combat related decisions. The DSS could assist the end-user in :-

(a) Assessing a situation.

(b) Planning operations.

(c) Enumerate alternatives.

(d) Evaluate alternatives.

(e) Simulate the execution of a chosen military action for having a feel of the ensuing combat and being ready for the various milestones in a pro-active manner.

(f) Making weapon selection decisions with computer aided techniques in real-time.

(g) Weapon guidance, where manual skills alone are not enough.

(h) Logistic support planning, automatic replenishment decisions.

(j) Resource for personnel tracking to assist the commander in knowing what he has at that time.

Some real systems, on which the information is available are briefly described below to elaborate on the "Decision Support System" applications, as enumerated above. But due to security constraints, the most important decision support systems are also the most zealously guarded technological secrets. Hence we will try to describe the systems, as known from the snippets of revelations made from time to time.

Command Planning Systems

These systems are the threat evaluation and resource management systems at the top most level. For example, in the case of the US, these systems will find use in the 'Pentagon' or Command Headquarters. These systems are used to assist political decision making of GO/NO GO kind or the military alternatives of "When to do, what and where". These systems are custom-built and but for their existence, little is known about them. They assist in :-

(a)   Quantification of threat , in terms of the enemy force.

(b)   Estimation of the Military Effort required , in terms of the availability of own forces under the following categories :-
- Strategic.
- Mobile Ground Forces.
- Air Forces.
- Naval Systems.
- Logistics.
- Transport Support.
- Re-deployment of Space Based Systems.

(c)   Operation Plans.

After the analysis mentioned in paras (a) and (b) are over, the estimated time for the deployment of friendly forces is estimated by interplaying the total transportation requirement with factors like transport support available, acclimatizing time, preparation time at new location etc. Thus the "time on battle front (D-Day)" can then be predicted. The battle-plan takes off from this starting point. The battle plan consists of a chronology of planned set of offensive/defensive manoeuvres, which will be conducted to achieve the desired effect on the enemy. The battle plan also has a modular structure. The aim of the operations translates into the macro plan, the macro plan is divided into sub-macro modules which in turn are split into micro operating plans. The amount of command/feed-back/coordination information can be gauged from the fact that planning at each of these levels (i.e macro, sub-macro, micro) has to take into account all the types of forces, i.e, air, ground, strategic, offensive, defensive etc. and their co-ordination requirements. A decision support system for assisting operations planning should, therefore, be capable of:

(a) Taking in data of friendly forces and their disposition.
(b) Taking in data on enemy forces and their disposition.
(c) Accumulate a library of friendly and enemy military system capabilities and their interactions/ comparisons.
(d) Should have models to compute efficacy of weapon systems or deployment under various scenarios.
(e) Should have a repertoire of practices/strategies/ tactics for operations and a set of validation criteria of the type " Must not do/May not do/Desirable /Must do". These two tools are the basic facilities for plan formulation.
(f) If possible, should have a presentation system in graphics mode to simulate the battle field view.

The Planning Decision Support System will provide the user:

(a) Present Location of Resources.
(b) Best Transportation Plan.
(c) Optimized Deployment Plan.
(d) Optimized Operations Plan.
(e) Information about the critical events so as to give a pointer towards success probability factor.
(f) A pointer to the suitable command and control structure for the operations in the battlefield.

Worldwide Military Command and Control System (WWMCCS)

It is a network of sensors, information systems and communication networks to be used by the National Command Authorities (NCA), the Chairman of the Joint Chiefs of Staff and the Commanders-in-Chief (C-in-Cs) to control the US forces throughout the world in peace time, crisis or war. It uses over 80 different satellites, radio and land line communication systems into some 30 command centres throughout the world. WWMCCS (pronounced as 'Wimmicks') aggregates all intelligence and status data on the US and likely enemy forces and is also used to convey command information back to the US forces. The decision support facilities in WWMCCS have been growing all the time and in 1982, Wimmicks Information System (WIS) programme has been initiated to update the C^3I structure as well as decision support utilities of the system.

Data Fusion System

TRW's Data Correlation (DACORR II) System is a data fusion system built on the state-of-the-art artificial intelligence (expert system) technologies. It is graphics oriented, interactive and has knowledge-based architecture. This architecture permits one to incorporate heuristic (rule-based) decision making. The system supports army commanders to marshall their forces for the optimum deployment in the evolving scenario of battle. The system has been used to manoeuvre battalions to echelons at theatre level and above with the following decision aid capabilities :-

(a) Situation Assessment.
(b) Critical Combat Node Development.
(c) Situation Projection.
(d) Combat Manoeuvreing Planning.
(e) Fire Support Targeting.
(f) Intelligence/Electronic Warfare Collection Management.

The DACORR permits simultaneous assessment of numerous hypothetical battlefield scenarios, providing a weighted assessment based on actual situation finding. The system uses advanced artificial intelligence and software concepts to provide a really powerful decision making tool in the battlefield.

Advanced Tactical Command Centre (ATACC)

A system being developed for US Marine Corps, for planning and directing co-ordinated air operations in support of ground units. It provides decision support facilities for force management, battle management, mission planning and air-task order generation. In the field, the ATACC will automatically generate air task orders and provide decision support, communication processing, graphics and text displays, information management and monitoring of friendly and hostile forces.

Battle details such as overall defence situation, enemy threat, and unit readiness will be communicated instantly to the air commander on ATACC automated displays. The commander will be able to issue orders immediately over digital and voice communication links. Monitoring and transmitting from a selection of 40 radios is provided by a touch screen radio panel. A single touch on the panel gives access to all ATACC radios in broadcast mode.

The wide range of information that can be displayed on high resolution displays includes weapon engagement conditions, classifications and air defence alert conditions, latitude and longitude data, time in local and Greenwich zones, alerting cues and universal transverse mercator (UTM) co-ordinates, which are used by pilots to plan routes. The system thus provides a decision support facility for almost all the facets of marine operations, instantly and comprehensively.

All Sources Analysis System/Enemy Situation Correlation Element (ASAS/ENSCE)

Information is to be gathered from many sources for war purpose and then processed to generate a total battle field picture with all the elements of enemy located in it. The ASAS/ ENSCE system is intended to automate intelligence function by collecting data from multiple sources, evaluating and correlating it, then disseminating a common, coherent, near real- time battlefield picture to airland battle managers.

The concept of the system is to allow military commanders to take decisions in 'near certainty' knowledge of the scenario. The concept emphasizes rapidly moving friendly forces, heavy use of Electronic Warfare and delivery of high speed weapons under computer control, often beyond visual range. The system design provides the real-time performance that will allow the commanders 'To turn within the Enemy's decision loop' i.e, before the enemy can detect the change in scenario and tackle it, the commander would already take the benefit of the scenario, or would have changed it in such a way that enemy reaction would have become inappropriate. This system will give the commanders not only the overview of the whole battlefield, but also allow him to monitor enemy's actions and reactions. This decision support system has the capacity to make the battle one-sided in favor of the forces who use it against an enemy who does not.

Manoeuvre Control System (MCS)

The objective of MCS is to allow commanders to work inside their opposition's decision cycle, i.e, the ability to change one's course of action before the enemy can change his. To attain this objective, MCS will automate command and control functions in armour, infantry and combined-arms formations. The system will also interface with other command and control systems related to fire support, intelligence, electronic warfare, air-defence and combat service support.

The equipment consists of a network of computers, which enable command and staff elements in the brigades, divisions and corps to interact with each other in collecting, analyzing, generating and distributing tactical information and directives. The facility allows total coordination among the battlefield forces and in addition allows front-end formations to view/review their operations in light of emplacements of friendly/enemy forces.

The description of the Decision Support Systems can be endless, as every facet of combat is now attempted to be provided with DSS so as to make the decision maker aware of the scenario, allow him to automatically enumerate (with assistance of computer memory, of course) the various alternatives and to help him to evaluate the alternatives and choose the best one. Thus the usage of DSS allows a combatant to take information based decision & take faster decisions.

Hence, each facet of the combat is a potential candidate for application of Decision Support Systems. There is absolutely no doubt, that an adversary having the DSSs in its inventory will have an overwhelming advantage. Gulf war is a conclusive proof of this assertion.

The components have been described with an aim to convey to the reader, the complexity, necessity and the hi-technology involved in the management of information during the war. All these components of information management systems had to work as one integrated real time system, for which necessary software, computers and structural organisation was provided to Schwarzkopf. A large amount of developmental work was still required for the system to work in the new environment of the MNF on one hand and Iraq on the other.

Information Age

The whole purpose of these dense, secure, jam-resistant and electromagnetically compatible communication systems was to let the messages pass amongst hundreds of users through different types of media such as satellites and various radio trans-receivers etc. As the flying intensity was very high, the reliability and speed of message transmission was also important. The information pertaining to navigation, meteorology, maintenance and administration and, the most important, for operations, had to be made available to the right person at the right time and place. This information management was another herculean task which was done commendably.

Maintenance : Quiet Heroes Behind the Theatre

The right targets were attacked at the right time. All this could not have been possible without near perfect and reliable maintenance of all the aircrafts under the most demanding conditions. To execute the Daily Task Order, the thousands of aircrafts, detailed for sorties, had to be well maintained and completely reliable. If a few aircrafts misbehave, not only a target would be missed, a vicious chain reaction could start as the next attack sortie may depend upon the performance of the previous one. To give just one example, if a particular radar or SAM station is to be destroyed before an attack mission can safely fly over that area for a target ahead, a failure of the radar destruction mission would jeopardise the passage of the next attack mission and, if attempted, that attack mission may go waste and so on. When during peak time, 2-3 aircrafts are to take-off every minute, the time slot allotted to a particular mission of aircraft would go waste and sometimes it may create confusion resulting into wastage of further sorties. Even if aircraft for that mission get ready to take-off after some delay, the complete sequence of events or allotment of aircrafts have to be immediately modified and all concerned need to be informed.

Frequency Allocation

The credit of executing such a large number of complex, difficult and interdependent sorties goes also to those who designed the high-tech computer systems (both hardware and software), communication systems as well as those who planned it with a high degree of precision. It was important that these communication systems worked flawlessly as their failure would have had a very large scale effect. Not only the dense communication system was required to function at thousands of radio frequencies in the e.m. spectrum, jamming of hundreds, if not thousands, of enemy frequencies was also being done and yet the communication systems of the MNF worked effectively. At times, a friendly transmission may unintentionally interfere with other friendly communications. It only means that frequency allocation is not as simple a task as it may appear to an uninitiated and were near perfect and the operators were disciplined and efficient. All the users had to sit together to decide and approve the frequency allocation plan. This is a very complex exercise involving a large number of dynamic variables and parameters with limited number of e.m. frequencies being available.

Compatibility Leads to Happy Marriage

The fundamental rules and regulations for frequency allocation and planning of their use are made during peace time, as far as possible. In the USA this task is carried out by the Joint Staff Committee (a combination of all the four forces), Defence

Communication Agency, Joint Tactical Policy Command, Control and Communication Agency and Military Communication Electronics Board. For allies, an additional Joint Communication Board carries out this task. This board consists of The Director of Command, Control and Communication and Computer Systems of USA and the member representatives of UK, Canada, Australia and New Zealand. In the management of this task, 'Electro Magnetic Compatibility (EMC) Analysis Centre' of the Department of Defence also assists. In codes and cipher management, representatives from Military Communication Electronics Board and National Security Agency are also involved. Computers and special communication systems were also developed for this purpose. For example, Tri-services Tactical Communication Systems (it contained circuit switches and message switches), lap-top computers, computer-computer interfaces were developed.

Joint Chief of Staff, General Collin Powel had stated, "Till now, command and control have been the key to success in war and will remain so whenever we get involved in a war." Such a detailed and tight command and control has been made possible by high-tech communication and ECCM systems. Thus, if at all a scheme of flying 2000-3000 sorties a day could be conceived, it is because of the use of high-technology products like computers and information management systems as well as their proper management, deployment and execution by the MNF; and their non- availability (due to ECM) to Iraqi forces. It is also a fact that electromagnetic interference from systems such as AWACS, jammer pods and airport radar caused computer misbehaviour, probably due to which 24 Patriots were launched at non-existing targets. Such a behaviour was exception rather than a rule and therefore the task of frequency allocation, without any doubt, was done extremely well.

RPV Tells What is Behind That Hill

'What is behind that hill ?', is a question that always worries a soldier during a war. Without knowing the strength and plans of the enemy, it is always dangerous to march ahead. Answer to this question can safely be provided by a remotely piloted vehicle (RPV) or with a low cost pilotless aircraft (PLA). With the help of appropriate surveillance and reconnaissance systems, RPVs can gather photographs of radar systems and troops movements, IR photographs etc. of the enemy troops and can gather ELINT (Electronic Intelligence). RPVs can also transmit the data in real time through a communication channel or the data recorded can be retrieved along with the RPV systems. RPVs can also illuminate targets with laser beams, whenever required for the laser guided weapons. RPVs have now been developed which can be used as ECM platforms and some of them, like anti-radar missiles (ARM), can be made to carry out a suicidal attack on radar systems. In Kuwait War, the MNF had successfully used 6 Pioneer and 4 Pointer RPVs for some of the difficult tasks stated above and thus have given a brief glimpse of the future wars.

It is surprising that in such an expensive war only a few RPVs were used. It is also not easy to understand, why the progress of development of RPVs has been slow over the years, specially in this high-tech era which has produced a missile like Tomahawk. Is there a bias against RPVs ? Just as there was a bias against air power in the early stages. Is it possible that 'pilots may have a bias against RPVs because RPVs provide a competition to the pilots. Or is it that they are too expensive, for the information they provide ? Or now there are more reliable and more resourceful options like satellites and special aircraft which are doing a satisfactory job ? RPVs are systems that need to be studied with an eye on future.

How is The Weather - As Fickle As Ever

Highest Technology is generally required for space systems. In this war, MNF had used the best available space systems. For air attacks, knowledge of present and future weather information is essential otherwise both time and sorties are wasted. The weather during the war period turned out to be the worst in the last 14 years and yet thousands of sorties could effectively be flown every day, thanks to the high-tech weather satellite pictures. Apart from the weather satellites, the USA had global positioning system and various surveillance and communication satellites which had helped in safe flying.

Global Positioning System : Where in Hell Are You ?

The global positioning system consists of 28 (21+7) satellites orbiting the earth at various levels at a height of 16,000 km. During Kuwait War there were only 13 such satellites orbiting the earth but even then the system proved to be of extreme help in navigating in that featureless terrain of the desert and during nights. This GPS system was extensively used by soldiers, convoys, Pumas (search and rescue helicopters), and SLAM missiles etc. Many lives were saved because at critical junctures no time was lost in loeating the victims to be rescued.

Surveillance Satellites : Big Brother Is Watching You

USA had four types of surveillance satellites viz. 3 electro-optical satellites (KH-11), 3 IR sensing satellites (advanced KH-11) and electro-magnetic satellites, 3 signals intelligence (SIGINT) satellites and 1 'La Cross' radar satellite. The KH-11 sends photographs immediately by electro- optical imaging whereas the advanced KH-11 uses IR imaging and is more useful during night time. They have a resolution of 15 cm i.e. a capability to discriminate two objects spaced merely 15 cm apart. 'La Cross' satellite sends radar imagery and has a resolution of 0.61 to 3.05 m. The SIGINT satellites receive messages on all the important channels assigned to them, sort out the messages, and transmit them to the designated earth station. Under the 'Defence Support' programme of Space Command of US Air Force, two missile warning satellites had conveyed the three dimensional position of Scud launchers, within 2 minutes of their launch, which the Space Command relayed to the Patriot missile units in the Gulf area. Ultimately, the Air Force Space Command had succeeded in establishing a direct communication link between the missile warning satellites and Patriot batteries of army, which used to get 5 minutes warning of the Scud missile's attack which was just about sufficient for

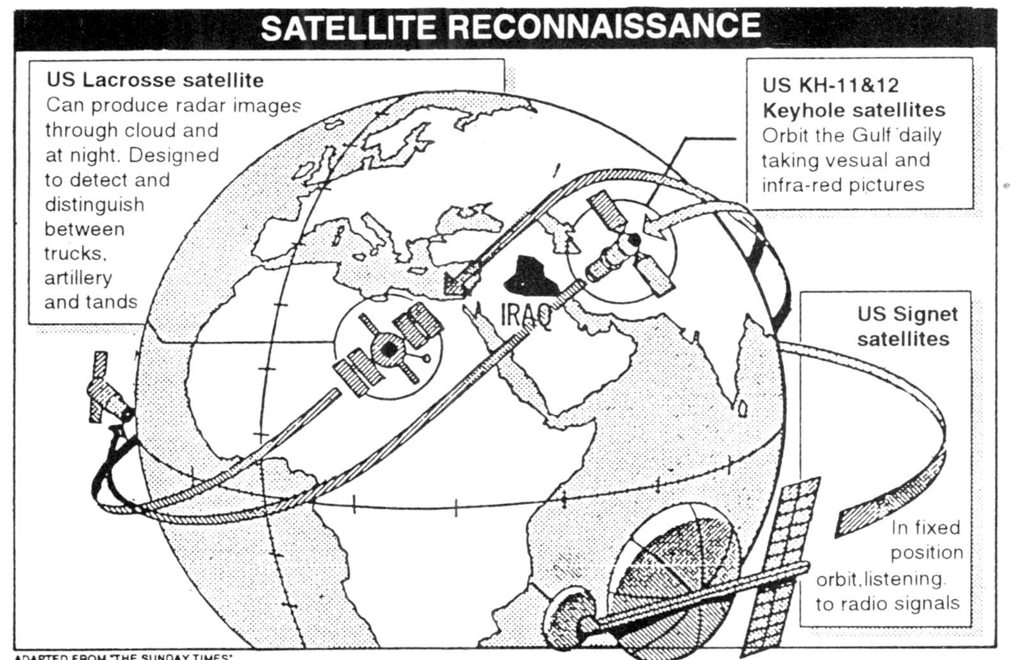

Figure 23
Satellite Reconnaissance

patriot missiles to intercept the Scuds. Narrunger, in South Australia, is a combined Space Intelligence Station which also helped in relaying the messages about Scud locations. The US and Australian Joint Naval Communication Station also served as a major communication link between different US aircrafts. The surveillance satellites gave the information of changing enemy orbats almost instantly as well as conveyed the extent of target destructions and battle damages of the enemy, in real time. It is only a simple example of information management during the Kuwait war.

Fairchild of USA and Spot Corporation of France sold mission support system and digital photos, respectively, to the MNF. The Commander of US Air Force Space Command, Lt. Gen. Moorman said that in Kuwait War it was for the first time that all the space based resources were really used and it would give a new direction to the history of war.

Command, Control, Communication And Intelligence (C^3I)

Strategy is the art of fighting battles to win a war. Therefore, there has to be an operational control centre or command centre where these battles must be planned and the battle commanders are given objectives and goals as per the plan to win the war. Therefore, all requisite intelligence must be made available to the Commander-in-Chief. Requisite communication is to be provided so that necessary command and control can be exercised appropriately. These four elements viz. command, control, communication and intelligence form the brain centre of the war. For instructions and decisions to flow-out and requisite information and feedback to flow-in, an appropriate hi-tech information management system is essential in the modern, fast, expensive, dense and complex war. In a modern war, therefore, the prime enemy targets are the enemy's C^3I centres; in order to enable the air power to attack these, air superiority has to be created in advance with the help of electronic warfare, higher/ attack aircrafts and surface-to-surface missiles.

"Once it was important that air battles were fought before commencing any action on a battle front during a war. But, it is my beleif that now a battle against radars is most essential, on which the success of air battle depends".

> \- Neil Cameron,
> Chief of the Air Staff,
> Royal Air Force (Early Forties)

As technology has advanced since WW-II, we need to amend the statement by Neil Cameran as "... now a battle against military information centres (radars are a part of it) is most essential..."

> \- V.M.T.

"War is Peace, Freedom is Slavery, Ignorance is Strength".
- George Orwell: '1984'(novel).

Chapter 6

Preparations for the Gulf War 91

Shielding The Desert : Preparations By MNF

Iraq's attack on Kuwait on 2 August 90 galvanized the USA into action. On 8 August 90, the US announced their first operation - 'the Operation Desert Shield' - implying defence of Saudi Arabia. Actually preparations were made to liberate Kuwait, if necessary by attacking Iraq. Whether you agree with the views of the USA or not, you would agree that their preparations were commendable, par excellence and exemplary.

The decision to pursue the military option was taken by US and NATO countries immediately after the occupation of Kuwait. Though, the exercise of the military option was still many months away, as diplomatic efforts were on way to force Iraq to vacate its military occupation of Kuwait, US systems were beginning to be manoeuvred into position for information gathering. Once again, it is clarified that the description which follows has been neither necessarily propagated nor denied by the US Department of Defence; and has been either garnered from snippets of press briefs during the Gulf war or has been inferred by analysis from the events of Gulf war.

Preparations for Information Gathering Exercise: The following activities were initiated for information collection during the war (The importance of 'information' in planning the war, before as well as during the war, would be clear by the fact that so many hi-tech techniques were used):

(a) The surveillance satellites were re-configured with photographic cameras and Infra-red sensors to locate Iraq's key strategic installations like command and control centres, oil installations, communication stations, army formations, bridges, nuclear facilities, factories supposedly involved in manufacturing chemical and biological weapons, airbases and missile units.

(b) The electronic surveillance satellites were re-configured to monitor radar and communication emissions to find out the locations of the emitters as well as the parameters of these emissions.

(c) The weather satellites were re-configured to monitor round-the-clock meteorological data for the battle area.

(d) At the Pentagon, at highest level, the military threat from Iraq was estimated qualitatively and quantitatively. Also, the Decision Support Systems (DSS) were used to enumerate various force combinations to counter it. It seems, that the US would have gone it alone, if the need arose, and on this assumption, the total force estimation was done. Later on, when other nations joined the military effort, the equivalent US effort was removed so as to leave the total force levels, as computed earlier.

(e) The force structure finally chosen shows that a lot of deliberations, enumeration and evaluation of many alternatives had taken place before the force levels were decided.

(f) The next problem to be solved was staffing. Having decided the types and quanta of weapon systems, combatants and support systems, the plans had to be translated into operational orders. The operational plans included identification of units, their place of deployment, their mode of movement and the training to be imparted before deployment. The extensive database of inventory and personnel plans assisted in mobilising the required forces. It is estimated that the first units, comprising 10% of the force, were 'combat-ready' within a month, 50% within two months and the rest within four months. The scope of information included army, air force, naval, marine and strategic elements.

(g) Before the transported soldiers could be taken as 'combat-ready, an extensive command and control network, intelligence and communication infrastructure and requisite information system/decision support systems were established at required points and centres.

(h) The final step before jump-off was to train the combatants to acclimatise them to the operating environment and also make them practice operations with deployed weapon systems, command and control infrastructure and other support systems. Some of the combatants, being from other nations, their initiation into intelligence, C^3I and DSS systems was a necessity to ensure co-ordination and control.

(j) The actual operation began after the diplomacy failed to dislodge Saddam from Kuwait. The war was predominantly an air-war (except for the 'formality' of last 100 hours), is also demonstrated by the pre-eminence of the information systems for air operations. That the air war became totally one-sided , the credit goes to electronic warfare, hi-tech weapon systems and C^3I systems. A significant part of the limelight must, however, be shared by the intelligence systems and their inherent real-time communication, computation and decision making support to hi-tech weapons, which made the Gulf war air operations so very successful.

(k) Discussing Gulf war will be an incomplete exercise without talking about Scuds. Though, in the final analysis, these missiles did not do much , but their terror value as probable carriers of weapons of mass destruction was high till the end of the war. The way Scuds were intercepted is fascinating and well publicized. What is not well known is "Scud-hunting" i.e. destruction of Scud sites (launchers) by F-15E aircraft. In any case, both these activities would never have been possible without extensive information and automatic decision support systems.

(l) The information systems played an important role in damage assessment, situation correlation and re-assessment, and operations management. In fact, the air operations were so well controlled that the relentless drubbing forced Iraqi soldiers into abject surrender. The situation assessment and event sequencing was so perfect that Iraqis could never use their armour. General Schwarzkopf and his team deserve all the accolades they got as commander of the MNF forces, but electronic warfare and C^3I systems deserve as much, if not more.

J-STARS Rise Early : The US Air Force and Army both were very keen to develop a special aircraft, the time schedule for which was 50 months due to the complexity of the project. In September 90, General Schwarzkopf sent a demand for two such aircraft. The development activity was expedited and on 13 January 91, these aircraft along with the aircrew, ground crew and maintenance personnel reached Saudi Arabia. These two aircraft were J-STARS - Joint Surveillance Target Attack Radar System. Such an aircraft is designed to detect low altitude slow flying aircrafts, helicopters, sailing ships, wheeled and tracked vehicles within a range of 200 Km. During a battle, if we know the strength and movement of the enemy forces without letting them know that of ours, it could be taken as a major step towards victory. With the knowledge of the strength and movement of enemy tanks, their assisting force, their communication set up, their order of battle, the attack or defence operations can be planned suitably and also one can attack with an element of surprise at their most vulnerable point. As it turned out, this war remained one sided, part of the credit must go to the efficiency, effectiveness and the punch with which the attacks were carried out which further demonstrate the utility of systems like J-STARS. Just as AWACS carries out electronic and radar reconnaissance for vitally important air attacks, JOINT STARS carries out similar tasks for land battles. Such a system not only helps in winning the battle, it also results in considerable economy because it enables defence or attacks at the right time with the right weapon on the right target.

Wild Weasels Made Smart : Prior to the Kuwait war, the Wild Weasels always moved in pairs on attack missions, one aircraft to search and precisely locate the victim radar by using its direction finder and the second, to launch its weapons, bombs or rockets. Now with the availability of High Speed Anti Radiation Missile (HARM) and a superior Radar Warning Receiver (RWR), both products of hi-tech, it was decided that only one aircraft should do this task. The attack aircraft, F-4E, of the earlier team was dropped and the search aircraft, F-4G, was modified to launch the HARMs. The RWR APR-47 was improved with a digital signal processor to enable it to identify the victim radars in a dense electronic environment and provide homing with an incredible accuracy of about 1^0. These modifications were also carried out ahead of schedule to meet the impending war requirements.

Patriots Upgraded : The threat from Scud, the Iraqi surface-to- surface missile, was also to be met. It was also feared that Iraq may use chemical warheads in its Scuds and hence a suitable anti-missile missile was essential. To meet this contingency, a surface-to-air missile system, Patriot, was hurriedly chosen to be modified. This upgradation programme was completed by December 90, which normally would have commenced by early 1991.

Patriot is a medium and high level SAM system which was originally designed for interception of fighter aircrafts. Soon after its successful trials, it was felt that it could possibly be upgraded into an anti-missile missile by improving its reaction time and accuracy. The breakthrough was possible because it had used phased array radars (AN/MPQ-53). In these radars, the e.m. beam of the antenna rotates by electronic devices unlike the previous generation radars which used mechanically rotating antennas to search the entire azimuth. This improved the reaction time of this SAM by an order because the radar beam can be made to rotate almost at any speed in any direction by electronic means. Further, it had used an e.m. frequency in a band of 4 to 6 GHz (1 Giga Hertz = 1000 Mega Hertz) which gave it better accuracy in azimuth and elevation.

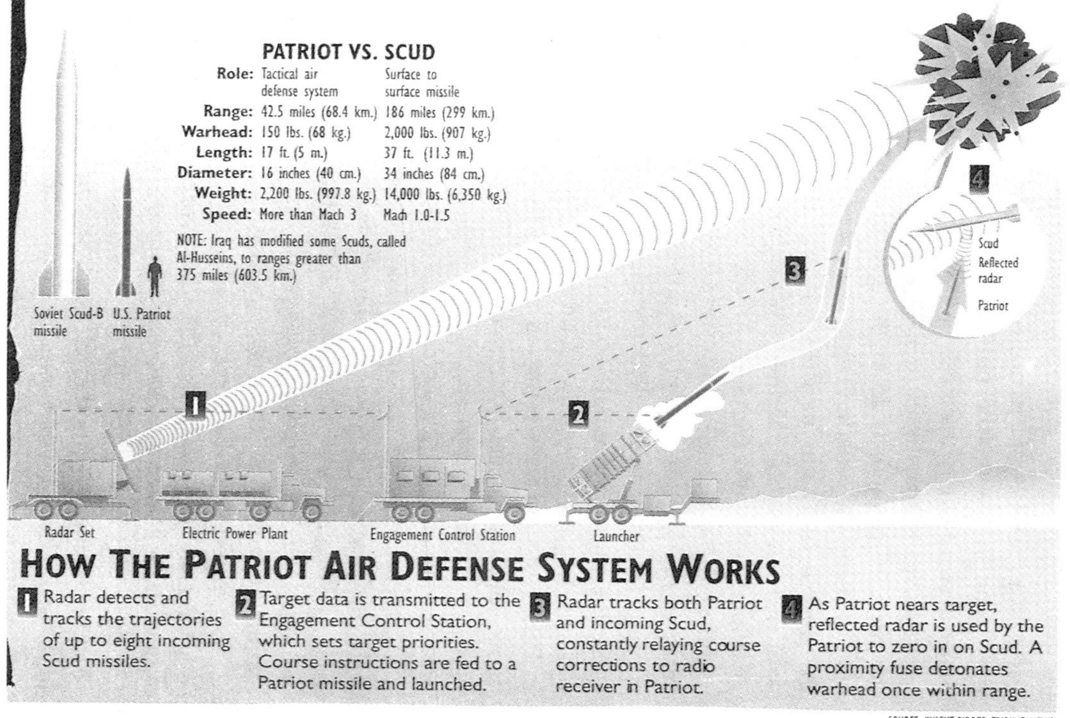
PATRIOT VS. SCUD

Role: Tactical air defense system | Surface to surface missile
Range: 42.5 miles (68.4 km.) | 186 miles (299 km.)
Warhead: 150 lbs. (68 kg.) | 2,000 lbs. (907 kg.)
Length: 17 ft. (5 m.) | 37 ft. (11.3 m.)
Diameter: 16 inches (40 cm.) | 34 inches (84 cm.)
Weight: 2,200 lbs. (997.8 kg.) | 14,000 lbs. (6,350 kg.)
Speed: More than Mach 3 | Mach 1.0-1.5

NOTE: Iraq has modified some Scuds, called Al-Husseins, to ranges greater than 375 miles (603.5 km.)

Soviet Scud-B missile
U.S. Patriot missile

Scud
Reflected radar
Patriot

Radar Set
Electric Power Plant
Engagement Control Station
Launcher

HOW THE PATRIOT AIR DEFENSE SYSTEM WORKS

1 Radar detects and tracks the trajectories of up to eight incoming Scud missiles.

2 Target data is transmitted to the Engagement Control Station, which sets target priorities. Course instructions are fed to a Patriot missile and launched.

3 Radar tracks both Patriot and incoming Scud, constantly relaying course corrections to radio receiver in Patriot.

4 As Patriot nears target, reflected radar is used by the Patriot to zero in on Scud. A proximity fuse detonates warhead once within range.

SOURCE: KNIGHT-RIDDER TRIBUNE NEWS

Figure 24

In order to improve the accuracy further, hence increasing the probability of kill, a modification was carried out in its semi-active approach. In a semi-active SAM missile, the transmission of the tracking beam to detect the enemy aircraft is done by means of a ground based transmitter. Reception of the return echo is done in the missile unlike in the case of 'command guidance system' where the receiver is also based on the ground and instructions to the missile on its journey to attack are passed in the form of commands from the ground. In a semi-active system, by using the reflected beam from the target, the receiver in the missile finds out its direction and then automatically tracks it without waiting for any commands from the ground. In the Patriot system, the receiver in the missile finds out the direction of the target (like in a semi-active system) but then passes it to the ground station. The ground station with better computing facilities, then calculates the position of the target and also the most appropriate chase-path to attack the missile more accurately and passes the commands to the Patriot missile. Thus Patriot system uses a mix of semi-active and command-guidance methods to get better jam-resistance, sensitivity and accuracy in its anti-missile-missile version. It has to be more accurate because when an aircraft is a target, explosion of a missile in proximity of the aircraft can damage the aircraft critically. In case of a missile as a target, the missile, which is very much smaller, less vulnerable and which travels much faster, must be hit directly, preferably on the 'head' The development work on Patriot system was on and, at normal pace, was expected to be completed by early 1991. With the announcement of 'Desert Shield', the work on it was so expedited that it was completed by end 90 and the Patriot missiles were deployed in Saudi Arabia and later on in Israel.

SLAM Slammed : Another air-to-surface missile, Stand-Off Land Attack Missile (SLAM) was to go through its operational testing in January 91. Its development was also expedited so much so that it was launched on the second day of the war, on 18 January 91, from the aircraft carrier John F. Kennedy from Red Sea. SLAM uses a thermal imaging device for homing and thus can be used in darkness. Its kill range is 90 Km and at such a long range,it has a incredible accuracy of about 2 m. The launching aircraft can fire it and forget it, thus increasing the safety of the launching aircraft. The guidance is carried out from another aircraft at a stand-off distance.

Better Search and Quicker Rescue

A special electronic system was also quickly installed in the AWACS which, it was claimed, had saved lives of air crew. The system probably had something to do with aircrew search and rescue system.

Mission Planning Computer Expedited : When an attack aircraft is required to penetrate into an enemy territory, its pilot has to plan its activity minute to minute and many times even second to second. This planning is complex and takes considerable time. Once the aircraft enters into enemy territory, due to tension and multiplicity of various tasks, both foreseen and unpredictable, to be performed, the pilot has little time to change the plan and make corrections, if required. All the actions have to be planned in minute details, that too, due to fast changing scenario, at the eleventh hour. It is also surprising that as electronic systems, especially computers, are entering into fighter aircrafts to take over certain actions of the pilot, his workload is increasing. It is mainly because increasing number of sensors and their ability to provide information to a pilot is also increasing and more and more response is expected from a fighter pilot. In modern scenario, the role of fighter pilot has increased so much that he might take 2-3 hours for planning and preparing for an attack sortie of 45 minutes. Although the use of computers in mission planning had begun well before the 'Desert Shield ', the progress achieved in development of computer programs just before and during the period of 'Desert Shield' was far more than that made during past few years.

Green Systems Launched : During a war, untried weapons are seldom used because their reliability or effectiveness is not exactly known and therefore their performance cannot be depended upon. In a war, which is already full of imponderables and unpredictables, there is very little chance for experimentation. Napoleon did not favour rifles because he did not want to take a chance with unreliable weapons. However, the new technology was given high importance during this Kuwait war. Although F-15E strike aircraft was using a powerful and smart ALQ-155 jamming system before this war, another improved jammer ALQ-184 was made ready just before the war. ALQ-184, a far superior system along with an improved receiver ALR-56 and chaff dispenser ALE-145 were also installed, thereby making the entire Active Self Protection Jammer (ASPJ) a totally new system and all the three sub-systems were integrated in F-15E strike aircraft in Saudi Arabia just before the war started. And jammers are the most critically important systems in the chain of attacks, specially so in the early stages of war.

Softwares of all the electronic systems, especially the electronic warfare systems were updated as per the situation. This was a time consuming task. Many aircrafts were modified to make them suitable for launching new weapons e.g. the attack aircrafts F-111 and Jaguar GR-1 etc. were modified for launching AIM-9M, the latest air-to-air combat missile. EC-130, a communication jamming aircraft, was also improved in a quick time, especially for identification and jamming of communications in the new war theatre, which is vital because the jamming denies the enemy an effective 'command and control'. It was ensured that the system should be able to differentiate between various types of communications. When it was amply clear that Saddam Hussein would not vacate Kuwait and the Multi National Forces had started organizing in Saudi Arabia, Britain also modified its Tornado F-3 fleet by application of radar absorbing material as well as installation of effective smart jammers and more powerful engines. Attack aircrafts were fitted with night-vision devices. Development program of anti-radiation missile ALARM was expedited so as to make it operational by January 91 instead of end-1991, etc. etc.. Similarly, French forces worked on war-footing and e.g. improved their Puma helicopter fleet by installing receivers for Global Positioning System (GPS). Thus the MNF worked with a clear plan, diligently and devotedly, to get prepared for a tough war. They obeyed the age old principle of war - 'the more you sweat during peace time , the less you bleed in war'.

Now, the question of using 'green' weapons in a treacherous war needs to be understood. No nation would like to experiment with lives of one's own soldiers in a war, least of all the USA - engaged in an alien war. It is the advancement in technology and total quality management which have made design, production and testing so realistic and dependable that the 'reliability' of the final product is exactly what one had aimed for. The risk is really negligible, provided the requisite quality control has been maintained.

Heat & Dust Burnishes MNF : It is well known universal truth that a system can be only as effective as its operator. With this in view, the training of air and ground crew was given due importance by USA and during Desert Storm they were trained thoroughly. The environment of a desert was alien and demanding to the MNF soldiers, therefore without getting acclimatized to the desert the MNF would have been only partially effective. Not only soldiers but machines also had to be tried and used and valuable experience gained on their reliability, maintainability and patterns of failures under the burning heat and dust of the desert. Many vital modifications were carried out on helicopters, electronic systems, aircrafts to improve their reliability in one of the most treacherous dusty and hot environments.

A well qualified American GI was further acclimatized and burnished into a 'professional soldier'. Desert can be dangerously hostile to a pilot also. To enter into a cockpit parked in hot sun is like entering into an oven. Therefore either aircrafts have to be parked in shade, cockpit cooled by pumping cool air or the pilot has to prepare himself to be cooked for 2 or 3 minutes. Once in the air, the flat and large stretches of sandy desert do not give him any clue of his whereabouts and he could even lose his sense of direction. Flying in such a monotonous and burning hot environment may make him prone to disorientation and fatigue. Even a light wind results into raising of dust in the atmosphere which reduces the visibility and obliterates the distinction between the land and the sky, which all may result in a crash specially so if a pilot is fatigued. One should not talk much about enervating heat, sand- storms, and paucity of water, as a soldier is expected to be prepared for all this in any case.

Desert Shield: A Full Dress Rehearsal : Every year NATO-Forces carry out an exercise 'Bright Star' in the Middle East and for 1990 it was executed for Gulf countries. The MNF carried out many exercises throughout the 'Desert Shield' and some of them were so realistic that pilots and soldiers, during the actual war, felt as if they were merely repeating those exercises. As Iraq had Mirage F-1 fighter aircraft and the related armaments, France after some dilly dallying gave Mirage F-1 and its armaments as well as the missile guidance details to the MNF for practice purposes. During this practice, the weaknesses of Mirage F-1 system were discovered and the knowledge of which enabled the MNF to develop appropriate tactics- both flying tactics and jamming tactics- for exploitation of those weaknesses.

Stealth aircraft also took part in gruelling exercises where about 85% of the total number of sorties were flown during nights. All other aircrafts, including the fuel tankers, carried out night flying practice as it was observed by the MNF that night capability of the Iraq Air Force was particularly weak and therefore most of their (MNF) attacks during initial week were confined to night time. All the exercises and preparations made indicate that MNF were not expecting 'air supremacy' as early as they really achieved it. Special flights were conducted to test the new systems and armaments. Thus, till 15 January 91, the MNF made sufficient preparations on war- footing. All the necessary maintenance spares, test systems, sub-assemblies etc. were stored and suitable plans were also made to procure them from USA within 48 hours of demand, an achievement to be proud of! Strategies and tactics were also thoughtfully planned. And in the wee hours of 17 January 91, the MNF launched a well prepared offensive with a bang and the 'Desert Shield' got metamorphosed into 'Desert Storm'. What were the basic strategies and tactics of the MNF ?

Figure 25

Royal Saudi Land Force Avibras Tectran Rocket Launchers fire during a desert shield live-fire exercise in the Northern Saudi Arabian desert

Preparations by Iraq

On the other side, Iraq, the aggressor who had already occupied Kuwait, also prepared for the war. Their aircrafts made practice flights, but mostly during day time and that too merely about 100 sorties a day as against the thousands by the MNF. It was not because of lack of will, but partially because of Saddam's feeling that the USA would not attack, partially because 'defence' and not offense was the guiding principle of his concept of this war and partially because he may be conserving the spare parts of imported aircrafts. High technology was used to make bunkers extra strong, and hardened shelters were made not only in Iraq but in Kuwait as well. It organized its air control system comprising radars, communication centers and SAM units in an effective way. It virtually created a thick wall of SAMs along the border which could be penetrated only by suffering a high attrition rate. At that time it appeared to be a correct action on the part of Saddam to enable him to survive a surprise attack by the MNF. He had put his aircrafts in vast underground high-tech shelters. Various command, control and communication centres, which serve the purpose of a brain during war, were housed in suitably hardened bunkers to ensure their safety. It appears that the strategy was to reduce the damage from attacks, rather than fight, thereby to prolong the war with the hope of repeating 'Vietnam'.

Army the Dominant Arm : Iraq's main force is army. As it has very little sea coast, her Navy is small. From the political point of view, Saddam did not believe that air force can be an independent and equally important arm. Conceptually he felt that in any war an air force has a secondary role to play i.e. mainly to give a support to the army. Therefore, major preparations were made only for the Iraqi Army. He had deployed 5.5 lakh soldiers in Kuwait for its defence and about 1.5 lakh Republican Guards, the best Iraqi soldiers, were kept in reserve at Basra. It is also difficult to understand why was the best portion of an army kept in reserve ! It again indicates, Saddam had no serious intention of fighting a full blooded war.

Prolong the war with as little destruction as possible must have been his motto, and therefore deploy the best soldiers in the safest place.

Ditch-Cum-Bund : Attack Abandoned : In order to defend Kuwait, they laid mines in one-kilometer wide and hundreds of kilometers long strip all along the Kuwait-Saudi border and erected barbed wire fence. They also got trenches dug all along the border. Arrangements were made to fill them up with oil with the aim to put it on fire when enemy forces come near it. Bunds were so made that when a tank climbed on them, it exposed its soft belly which then could be attacked easily with devastating results. All these obstacles were not impossible to surmount but certainly very difficult ones as they reduced the speed of the attacking army and made them extremely vulnerable to losses. In such a situation the advancing army becomes an easy target to the pre- positioned artillery, machine guns and anti-tank weapons. Under these conditions the losses of an attacking army are heavy and therefore the requirement of the attacking forces is three to five times that of the defending forces.

In Kuwait the Iraqi Army was prepared solely for defence whereas the airspace depended for its safety mainly on the 'Air Control' - again a strong but defensive arrangement. Iraq had positioned 7 armoured divisions in Kuwait whereas the rest three divisions (Presidential Guards) were kept in Basra as reserve. Each of the armoured division was given 15 SA-6, 27 SA-7, 5 SA-8 and 16 SA-9 SAM systems and 16 Quad ZSU 23 radar controlled anti - aircraft artillery (AAA) for defence against air attacks. Such a formidable deployment of 63 SAMs and 16 Quad guns can prove suicidal to any attacking air force unless it comes specially prepared. For defending the airspace at low, medium and high levels, SA-2 to SA-14 Russian SAMs as well as some French SAMs (Rolland) were suitably deployed. All these systems had suitable 'fail-safe' communication systems.

What made Saddam either believe or not even think that the MNF would not attack from across the Saudi-Iraq border ? Firstly, Saddam had believed that the US would not attack at all, a suicidal belief because this prevented him for making realistic war like preparations. Secondly, the trenches were dug more to demonstrate to the US and Saudi Arabia that he had no intention of attacking Saudi Arabia, an ally of the US. Saddam really had made the defence on Kuwait-Saudi border so formidable as to be impenetrable, then he had left only two choices for attacking forces. Either they could attack from the sea side or from the Saudi-Iraqi border. Saddam prepared a defence for attack from sea side but he did not think of defending the Iraqi-Saudi border! Did Saddam think that an attack across the Iraqi-Saudi border would be a crime, a war crime ? Or was Saddam's mind-set that of a medieval war lord, who entrenches himself in his fort leaving everybody else at the mercy of raiders but with an aim to enervate the raiders ? The 'Fort' in this scenerio being 'Kuwait' which was protected by the 'walls' of SAMS and trenches. Out-flanking movements in wars are quite common, but every time they are carried out, they surprise the attacked party. WW I and WW II are full of such examples.

Thus Saddam had made an invincible (so Saddam had thought) looking fortress with his hi-tech bunkers, very dense deployment of SAMs and AAAs, ditch-cum-bunds, mines and a very large army which had just fought an eight year long war with Iran. In terms of hi-tech reality how invincible was this 'fortress' ?

> " The more you sweat in peace, the less you bleed in war, is also applicable to research and development of military technology including electronic warfare."
>
> V.M.T.

War is a valuable instrument of national policy in proportion as military techniques provide or enable the most economic methods for gaining results deemed important.

Chapter 7

Matching of Strengths

General Schwarzkopf was extremely well prepared by 15 January 91. He had the complete knowledge of the order of battle of Iraqi forces and he was continuously updating them. All the means of gathering intelligence, viz. cameras, IR sensors, electronic receivers ; and all the platforms with various modes of transport viz. trucks, aircrafts, ships, space-crafts, stationary units and human beings were used. Of course his 'human-int' was poor, as it was very difficult to get Iraqi agents, although he got some valuable help from the Kuwaiti Resistance Movement.

Multi-National Forces (MNF)

The Multi-National Forces (MNF) were constituted by the following 29 national forces :-

Argentina, Australia, Bahrain, Bangladesh, Belgium, Britain, Canada, Czechoslovakia, Denmark, Egypt, France, Germany, Greece, Italy, Kuwait, Morocco, the Netherlands, New Zealand, Nigeria, Norway, Oman, Pakistan, Poland, Qatar, Saudi Arabia, Senegal, Spain, the United Arab Emirates and the USA.

A Statistical Comparison

Military strengths of the MNF and Iraq are briefly given in the succeeding tables. Statistical information regarding complete war machinery was collected from various sources wherein, many a time, conflicting numbers were indicated. Due to these constraints, all attempts have been made to give only the authenticated figures , whereas for others the quantities have not been mentioned.

Force	Iraq		MNF	
Army strength	5,50,000	in Kuwait	5,40,000	USA
	1,50,000	Reserve in Basra	2,05,000	Others
	3,00,000	in Iraq		
	10,00,000	Total	7,45,000	Total
Divisons	7	Armoured	Since the MNF Army	
	42	Infantry	was drawn from	
	6	President's Guards	various countries, they	
		3 Armoured	could not be classified	
		1 Infantry	into Divisions and	
		1 Commando	Brigades etc.	
		1 Other		
Brigades	20	Special Brigades		
	2	Surface-to-Surface Missile Brigades		
Fire Power	5500	Main Battle Tanks	500	MIAI MBT,
	100	Light Tanks		M60AI,
	9000	Guns		Challanger,
	200	Multiple Rocket Launchers		Warriors M-110
			40	AMX 30,
				Chieftans,
	36	Stationery Launchers (Scud)		M 270,
				Multiple Rocket-
	50	Mobile Launchers (Scud)		Launchers
	50	Frog Launchers		
Anti-Tank Missiles		AT 3		Hellfire
		AT 4		Tow
		SS 11		Swing Fire
		Milan		Milan
		Hot		Hot
		Anti-Tank Guns		
Attack Helicopters (with their missiles)	160	BO-105 (with AS-11) MI-24 with AT-2		Apache AH-64 (Hellfire) Cobra AH-1

Force		Iraq		MNF
		SA-341		(Hellfire)
		SA-342		Black Hawk
		EXOCET		(Hellfire)
		AS-12		Lynx
				(Swing Fire)
				Gazelle
				(HOT)
				Puma
				(Milan)
Anti-aircraft		23 mm ZSU (Radar		20 mm Vulcan
Artillery		Controlled)		Guns
		23 mm Guns		25 mm Guns
		37 mm Guns		
		57 mm Guns		
		85 mm Guns		
		100 mm Guns		
		130 mm Guns		
	9000	Total		
Surface-to-	160	SA-2 Medium Level		Crotale,
Air Missiles	140	SA-3 Low Level		Cheperal,
		SA-6		Hawk,
		SA-7		Javelin,
		SA-9		Matra SA-10,
		SA-13		Mistral,
	300	SA-14	90	Patriot,
	100	Rolland		Rolland, Rapier,
				Stinger,
				Star Streak
Surface-to-		Scud B	300	Tomahawk
Surface		Al Hussein		C and D
Missiles		Al Abbas		(Minimum)
	400	Total		
Bomber	8	TU 22	48	B-52 G
Aircrafts	8	TU 16		
	4	H-6D		
	20	Toal		
Ground Attack	70	Mig-23 BN		A-6E Intruder
and Strike	64	Mirage F-1EQ		A-7E Corsair
Aircrafts	30	SU-7	180	A-10A Thunder bolt
	50	SU-20	80	AV-8B Harrier
	30	SU-25	48	F-15E Eagle
	40	J-6	249	F-16C Falcon
	20	SU-24	128	F/A-18

Force	Iraq		MNF	
			83	F-111
			56	F-117 A
			40	Jaguar GR 1A
			46	Tornado GR 1 and Tornado GR 1A
			27	Mirage F-1 CR
	304	Total	937+	Total
Air Superiority Fighter Aircrafts	25	Mig-25	114	F-14 Tomcat
	70	Mig-21	120	F-15C Eagle
	80	J-7	128	F/A - 18 Hornet
	30	Mirage F-1EQ		
	30	Mig-29		F-16 Falcon
	235	Total	18	Tornado F-3
			10	Mirage 2000
			390+	Total
Reconnaissance Aircrafts	8	Mig-25		RF-4C
			3	Nimrod
Early Warning Aircrafts	2	IL-76	10	E-3A Sentry
			30	E-2C Hawk Eye
			2	E-8 JSTARS
Electronic Attack Aircrafts	-		48	F-4G, Wild Weasel Tornado GR-1
ECM Aircrafts	-			EC-130 Compas Call
			30	EA-6B, Prowler
			18	EF-111 A, Raven
Elint Aircrafts	-			TR-1, RC-135 RF-4C, U-2R EP-3, EA-3B
Special Task Aircrafts	-		6	OV-10D Advanced Air Control
			50	HC-130 P, MC-130 E, and MH-53 J Buccanier
Fuel Tankers	-		256	KC-135
			17	VC-10
			46	KC-10
Transport Aircrafts	10	AN-2		C-130 E
	6	AN-12		C-130 H
	6	AN-24		C-5
	2	AN-26		
	19	IL-14		

Force	Iraq		MNF	
	17	IL-76		
	60	Total	145	Total
Total Number of Aircrafts	629	(653 ?)		About 2500
Air-to-Surface Missiles		AS-4		Penguin
		AS-7		Kormoran
		S-30 (Laser Guided)		AS-15
		AM-39		AS-30
		ARMAT (Anti-Radar)		Exocet
		Exocet (Anti-Ship)		AGM-45 Shrike
		C-601		AGM-65 Maverick
				AGM-142AHaveNap
				AGM-114 Hell Fire
				AGM-88 HARM
				SCUA
				Harpoon
				ALARM
				Swing Fire
				HOT
				MILAN
Air-to-Air		R-530 Matra Super		R-530 Matra Super
		R-550 Matra Magic		R-550 Matra Magic
		AA-2,		AIM-7 Sparrow
		AA-6,		AIM-9L
		AA-8,		AIM-9M
				Sky Flash
Naval Ships & their Weapons	5	Frigate, each with	7	Aircraft Carrier, each having 80 Fighter Aircrafts
		1 Anti-submarine Helicopter		F-14A or F/A-18,
		2 Anti-submarine Torpedo		A-6E,
		8 Surface-to-surface missile (Automat)		E-2C Hawk Eye, AV-8B, and Anti-submarine Helicopter
		1 127 mm Gun	100	Battle Ship Frigates, each with 1 Anti-submarine Helicopter 2 Anti-submarine Torpedo 8 Surface-to-surface missiles

Force		Iraq			MNF
					(Automat or Harpoon)
	38	Patrol & Gun Boat			Patrol & Gun Boat
	6	Corvet			
	6	Torpedo Vessels			Torpedo Vessels
	6	Amphibious Boat			Amphibious Boat
	20	Patrol Boat			
	8	Missile Boat			Missile Boat
	8	Mine Vessel			Mine Vessel

Force	Total	U.S.A.	U.K.	France	Others	
Army	7,45,000	5,40,000	33,000	11,500	47,000	Egypt
					19,000	Syria
					6,000	Bangladesh
					2,000	Pakistan
					1,500	Morocco
					480	Nigeria
					500	Senegl
					40,000	Saudi
					10,000	Bahrain
						Oman,
						UAE,
						Qatar &
						Kuwait
						etc.
Air Force						
Airmen	62,000	55,000	7,000	—	—	
Aircrafts	2,430 (16.01.91)	—	—	—	—	
	2,790 (24.02.41)	—	—	—	—	
Flights	1,08,000	1,00,000	6,100	1,879	—	
Armaments Used (Ton)	86,000	83,000	3,000	—	—	
Air Lift Strategic Equipment (Ton)	5,63,000	5,13,000	50,000	—	—	
Personnel (Air trans-ported)	—	4,82,000	—	—	—	

Figure 26

A U.S. Air Force F-16 fighter airplane is shown with weaponry it can carry - air-to-air and air-to-surface missiles, laser guided munitions, bombs of various sizes etc.

Figure 27

EF-111 A (Raven) on their ECM missions over the Saudi Desert.

Figure 28

F-111 Takes off from an air base in Saudi Arabia. For an Attack mission

Financial Assistance to MNF

The following 18 countries gave financial assistance on humanitarian grounds :-

Afghanistan, Austria, Bulgaria, Finland, Honduras, Hungary, Iceland, Japan, Luxembourg, Malaysia, Phillipines, Portugal, Sierra Leone, South Korea, Sweden, Taiwan, Turkey and USSR and United Nations (for UN, details given at the end of chapter3).

The Iraqi Army

Army occupies supreme position in the Defence Services of Al Jamhuriya Al Iraqia (meaning the nation whose roots are deep). Iraqi Navy is small as is its coast line. Iraqi Air Force operates under the control of their Army. A few years before this war the strength of its army was one million which is reported to have got reduced by 3 lakhs during the long Iraq-Iran war. In the beginning of February 91 it was published in newspapers that Iraq was recruiting youths of even 17 years of age. Iraqi army could have been sufficient to defend Kuwait if Iraq had not lost the air superiority and many other ifs. In comparison to the MNF, the Iraqi Army was superior not only in numbers, but in its fire power also.

The Iraqi Artillery

The Iraqi Howitzer (artillery guns) had a firing range significantly longer than those of the MNF. Consequently if the MNF guns were to shell Iraqi forces, they had to reach within the firing range of Iraqi guns whereas Iraqi guns could fire from a safe distance. However, it should be noted that such an advantage of artillery superiority would have been available to Iraq only if a war on the ground was really fought. In addition, Iraq had modern electronic fire control systems to measure the miss-distance of the shell and make corrections for the next fire. Such a system avoids the wasteful method of 'trial and error' and also obviates the necessity of an observer plane, which remains vulnerable to attack. Electronic systems are available which can convey the exact spot where the shell has fallen, on the basis of which suitable corrections could be made for the next fire. Further, the advantage of long range guns also necessitates air superiority, at least over the area of gun-deployment; otherwise enemy's aircraft can easily bomb the guns.

The Iraqi Missiles

Scud Missiles : Let us look at the capabilities of the Scud, the famous or in-famous surface-to-surface missile (SSM). It is an improved version of the SS-1 Soviet missile, and its technology is not far ahead of the technologies that were available to WW II weapons. Its kill range is about 300 km and its circular error of probability (CEP) is about 900 m, i.e. if 100 Scuds are fired on a target, only about 70 would fall within 900 m of the target. Such a CEP is rather poor as, after all, it is not a radar guided missile, and the gyroscope it uses to maintain its path is also of old technology. Although the weight of a Scud is 6370 Kg, its warhead is only 985 kg.

Al Hussein and Al Abbas : Iraq, with the assistance of Egypt, had produced two improved (?) versions of the Scud missile; Al Hussein, whose range is 600 km but the war head is reduced to 500 kg and CEP is increased to 1800 m and Al Abbas whose range has been further increased to 900 km but the CEP is also increased to about 3000 m. Thus it can be seen that all the three SSMs with conventional warheads were militarily rather ineffective and can cause unnecessarily large scale destruction only. These are area weapons and need to be fired in a very large number and therefore cause even larger unwanted destruction. To be somewhat effective, they should carry a nuclear warhead with a devastation range of 1 km so that if 2-3 Scuds are fired on a target would hopefully be destroyed, otherwise with conventional warhead the scuds can act only as a terror-weapon against a civilian population. Saddam used Scuds only as terror-weapons. A total of 81 of these were fired and they did not do any serious damage except killing about 25 persons including civilians. Attacks on the civilian population is a crime even in a war. With conventional warhead, Scud and its variations are, by

Range of Iraq's missiles
Iraqi Al-Hussein
Modified Scud-B
Range: 375 miles (can hit Riyadh or Dhahran from Kuwait)
Warhead: About 250 pounds
Chemical potential: Both Al-Hussein and Scud-B can carry chemical or nuclear warhead; if chemical, probably mustard gas
Soviet Scud-B
Range: 186 miles (unable to reach Saudi Arabian cities of Riyadh or Dhahran from Kuwait)
Warhead: About 2,200pounds; explosive power equal to one Stealth bomber
Length: 37 feet
Diameter: 34 inches
Weight: 14,000 pounds
Black Sea
Soviet Union
Al-Hussein range
Turkey
Istanbul
Iran
Med. Sea
Syria
Baghdad
Israel
Iraq
Jordan
Scud-B range
Kuwait
Egypt
Saudi Arabia
0   250
Miles

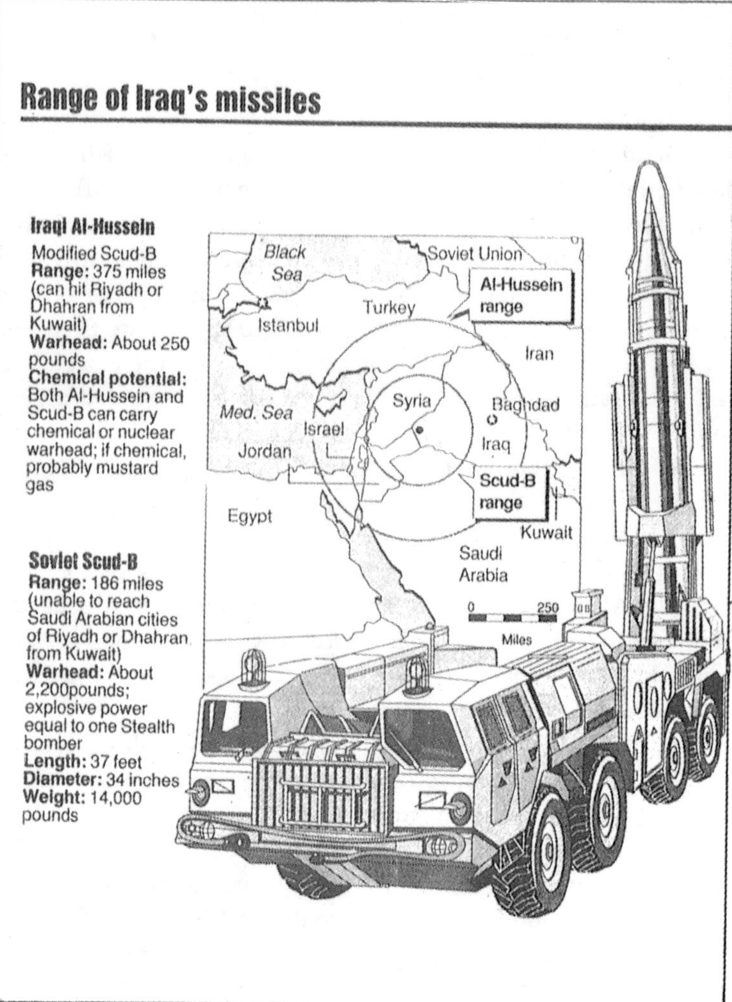

Figure 29

and large, terrorizing weapons, unless a military target with a large area is properly chosen.Hitler had also used V1 and V2 rockets as terrorising weapons by attacking London.

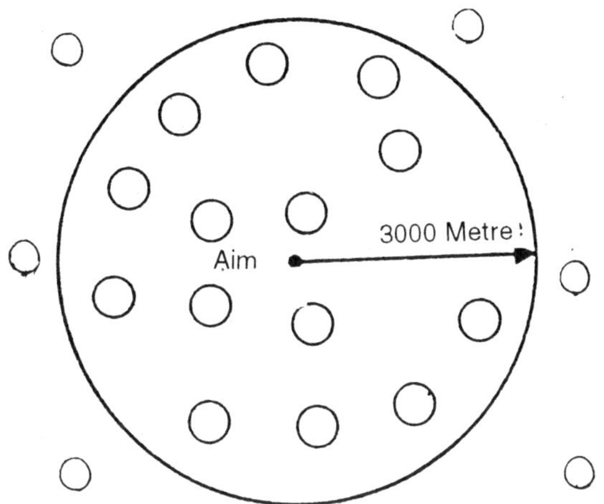

A CEP (Circular Error probability) of 3000 m means that if 20 Al Abbas SAMs are launched at a target in sucession , only 14 of them are likely to fall within a radius of 3000 m from the target.

Figure 30
Circular Error Probability of Al Abbas

A CEP (Circular Error Probability) of 3000 m means that if 20 Al Abbas SAMs are launched at a target in succession, only 14 of them are likely to fall within a radius of 3000 m from the target.

The Iraqi Aircrafts

Iraqi Bombers : Al Kuwwat Al Juvviya Al Iraqia (Iraqi Air Force) had a good number of long range and effective bombers like 8 TU-22s, 8 TU-16s and 4 H-6Ds (Chinese version of TU-16), all products of modern technology. Though they are much more modern than the US B-52G, but compared to same generation US bombers, they perhaps lag behind slightly in respect of modern electronic, navigation and weapon management systems for which the more backward US B-52G had been modified for this war.

However, the TUs could have still been used effectively only if they were used at the right time. TU-16 has an airborne radar which operates around the 'hi-tech' e.m. frequency of 14 GHz (1 GHz = 1000 MHz) with a range of 200 kms. It has modern ECCMs like frequency agility. TU-22, in addition to dropping bombs can also launch AS-4 air-to-surface missiles. AS-4 can guide itself with the help of an active radar and attack the enemy's radar up to a range of 300-400 km. It is a modern weapon system. Similarly TU- 16 can also launch AS-5 missiles which have a range of 300-400 km.

Iraqi Fighter Aircrafts : Iraq's air arsenal included the most modern MIG-29 and Mirage F-1-EQ series of fighter aircrafts. With the help of these aircrafts Iraq could have provided stiff resistance to the MNF, if not prevented them from establishing their air-superiority over Iraqi air-space despite latter's overwhelmingly large number. It could not have ultimately succeeded in preventing them from establishing the air-supremacy but it certainly could have exacted a price.

Mirage F-1-EQ-6 were produced till 1989-90 and delivered to Iraq in 1990. It is a multi-purpose aircraft and has the facility of air re-fuelling. It can launch modern air-to-ground laser guided missiles AS-30 and Exocet missiles etc. It is also armed with air-to-air missile 'Matra Super' which has 'look-up' and 'look-down' capability with a range of 40 km and it is a good match to the US AIM-7 Sparrow missile. The electronic systems of F-1-EQ like the navigation and weapon aiming computer, head-up display and inertial navigation system are product of modern technology and its radar has modern ECCM capabilities. Most of these aircrafts have modern radar warning receivers, IR missile warning systems as well as chaff and IR flare dispensers. The semi-active targetting radar of Matra Super also has good ECCM capability. F-1-EQ can also launch ARMAT, anti-radiation missiles, which were believed to be there in the possession of

Iraq. It has air-to-air missile like Matra Magic having 'close combat' capability. It had ECM systems like Barax, Baracuda, Caiman and Remora etc. Therefore, the Mirage F-1-EQ fighter aircraft is considered to be capable of handling air combat and ground attack operations in ECM environment. It is no surprise that the US pilots were concerned about Mirage F-1-EQ, before the war began and they made special preparations against it by obtaining, despite French unwillingness, all the necessary detailed information from France.

MIG-29 : MIG-29 is one of the best amongst the modern air superiority fighter aircrafts. There are four main characteristics that one should study while comparing fighter aircrafts. First is concerned with the basic flying capabilities of an aircraft like speed, power, acceleration, manoeuvreability etc. Second deals with its armaments like bombs, rockets, guns, missiles etc. Third covers the performance of the weapon control systems and their ability to launch the weapons under different conditions as well as in an ECM environment. Fourth is the man- machine relationship. In first three qualities, as per most experts, Mig-29 rates equal or even superior to the other two best air superiority fighter aircrafts of the US, F-15C & F/A-18.

The air-to-air missiles that MIG-29 can launch are AA-7 'Apex', AA-8 'Aphid', AA-9 'Amos', AA-10 'Alamo' and AA-11 'Anchor'. AA-8 and AA-11 are used for close air combat. AA-8 is a standard weapon, available in both the radar homing and the IR homing versions. AA-11 is superior to the best in its class, the US air-to-air missile AIM-9M, in four out of the five main qualities viz. range, warhead, manoeuvreability and off-bore sight launching. The fifth main quality is its ECCM capability, and AA 11 may not be superior in this quality to AIM-9M. Both the missiles, AA-8 and AA-11, have all-aspect capability, available only in the best of AA missiles, i.e. they can attack the target aircraft from any direction. The medium range AA-7 is comparable to AIM-7 Sparrow of the US. AA-9 'Amos'is the long range air-to-air missile and is comparable to AIM-54 'Phoenix' except that its maximum range is about 75 km compared to 150 km of AIM-54. AA-10B 'Alamo', also a long range missile, has IR homing system with 'fire and forget' capability, which is generally not found in long range air-to-air missiles.

As far as characteristics of control and launch of weapons are concerned, MIG-29 is unquestionably superior. It has an integrated effective weapon control radar system, consisting of electro-optical search and track system and helmet mounted target designator. Generally the electro-optical system searches and tracks the targets with a laser beam but maintains the electronic undetectability (electronic stealth), as it then does not use a radar. While a MIG-29 is carrying out such search and track operations, the target aircraft (let us take a difficult scenario), goes behind a cloud, the electro-optical system would lose the target but would then trigger the radar to take over the task, which would again go off as soon as the target would come out of the cloud cover and the electro-optical system would again takeover. Thus when the target is reached within the range of its missiles, the pilot has just to look straight into the target, the sensor in the missile would automatically lock-on to it (it reminds one of Lord Shiva's third eye). After this the pilot has just to press a button to launch the missile which would then chase the target. Such a well integrated weapon aiming and control system does not exist even in F-15C and F/A- 18A aircrafts.

MIG-29 : Human Engineering : In human engineering of man-machine relationship, the aircraft MIG-29 is weaker than comparable aircrafts. The pilot is over burdened with various operations, in addition, all decisions have to be taken instantly in an environment where time is always running out at supersonic speed. In such an environment, the man-machine design should be extremely pilot-friendly. In a fighter aircraft it is essential that the pilot should be able to handle all his armaments with his 'Hands On Throttle And Stick' (control column) - HOTAS. In MIG-29, the HOTAS management is not possible; further, the pilot's visibility in the rear direction is poor. If an aircraft, by maintaining electronic silence, approaches a MIG-29 from the rear and if the relevant ground radars are jammed, the enemy can come close enough and comfortably shoot MIG-29 or fire a missile. Cockpit arrangement is also not designed to give top priority to the pilot's convenience in mind. However, skilled pilots can overcome this weakness with extra practice and confidently fight with any air superiority fighter in the world.

Ground Attack And Strike Aircrafts : Ground attack and strike aircrafts of Iraq are also a mix of modern and not so modern technology. MIG-29 and F-1 EQ are multi-role aircrafts and therefore can effectively play the ground attack and strike role also. Of course they have to be provided with air-to-surface missiles, bombs and rockets etc. MIG-23 BN, swing-wing fighter-bomber aircraft is not so modern but it has a laser range finder which helps in hitting ground targets with precision, and it can launch, in self defence, air-to-air missile AA-7 'Apex'. MIG-25, though an older technology aircraft, could be modified with modern missile launchers to become a challenge to the enemy. It can launch AA-7 (range 40 km), AA-11 (a supreme close combat missile) and AS-11 (air-to-surface long range missile) and with little effort MIG 25 can suitably be modified for the 'Wild Weasel'role. It is not known what modifications, if any, were done by Iraq. SU-25 is another bomber they have, which is a match to American A-10 and can launch laser guided bombs. However, unlike SU-25, A-10 was used heavily in this war.

Factors in Air Combat

In any air combat speed and manoeuvreability is vitally important, as it is a question of life and death or victory and defeat. Air combat is engaged with an extremely expensive machine controlled by even more valuable pilot. An air combat is a highly complex and dynamic process which is full of complexity, split second decisions and tensions. Capabilities of fighter aircrafts along with skill, courage and morale of the pilot together decide the victor. But there is another extremely vital factor in an air operation, in general, and in air combat, in particular; and that is Air Defence Ground Environment System (ADGES) which is sine qua non for air defence of a nation. Main elements of ADGES are early warning radars, ground control interception (GCI) stations, three dimensional radars, acquisition radars, 'command, control and communication', air defence control centre and necessary interceptor fleet, SAM batteries and AAAs. All these elements function in an integrated network as shown in the tree diagram.

A Typical Air Defence Organisation Tree

All units shown in the tree must be provided with necessary anti-jam communication facilities so that virtually, second by second, progress is passed from one to the other. All communication channels and radars in ADGES must have sufficient jam-resistance facilities in them.

The organisation tree shows that it is hierarchical, and the information generally flows from bottom to top whereas the decisions and orders come from top to bottom. A hierarchical system has the advantage of using the resources very economically. The resources can be transferred from one to another, depending upon the need, and they do not have to be unnecessarily duplicated (practically, no one wants to give away his resources though he may not need them at that time). Further there are no command and control ambiguities. But the hierarchical system has the disadvantage of being too top dependent. Since every unit is trained and accustomed to take directions from the top, the lower echelons do not use their initiative and slowly it becomes a part of their habit. If one of the top units due to some reason becomes non-operational, the lower units become top-less and rudderless and the entire chain of command and control gets disrupted resulting in confusion. How much power, authority and responsibility is to be given to whom, depends upon the political, cultural and traditional behavioural pattern of the country and its defence forces. If the people of a nation are habitual of a hierarchical pattern, it is possible that they may still like sub-ordination, as far as decision making is concerned, even if independence is provided to them. They may continue to prefer taking directions from the top even on matters under their jurisdiction, and the superior officers also enjoy the extra authority and power, whenever it suits them. Appropriate freedom (with its consequent risks) and training at

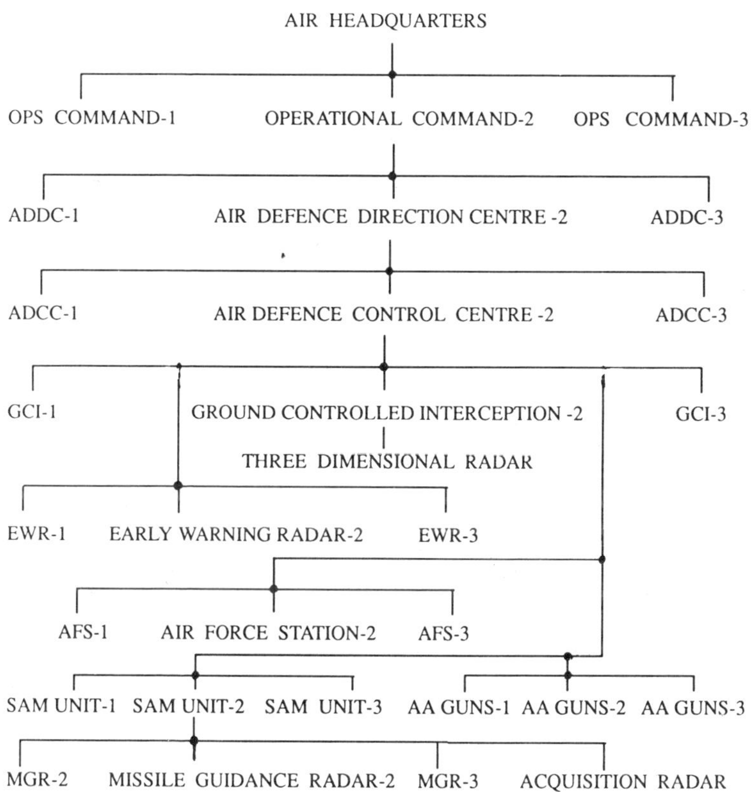

AIR DEFENCE SYSTEM ORGANISATION : AN HIERARCHICAL SETUP

Note : The chart gives only the conceptual idea of a typical field formation. Depending upon requirements, each country will have its own variations.

Figure 31

every level would make such an organisation for ADGES very responsive to dynamics of a war and that is what real leadership is all about. The organisation gets benefit of everybody's experience, ability, initiative and courage. Such a capability has to be inculcated with patience and dedication. Iraq has its Defence Headquarters and Air Force Headquarters in Baghdad and its three functional commands are in Kirkuk (North), Naseeria (South) and Rutba (West). Their air defence machinery was based on hierarchical pattern, with dictatorial culture, in which every level had very limited freedom.

Strength of the MNF : High Tech

The strength of a nation to wage a war depends not only on its weapon systems and soldiers but also on its information management system, practical organisational behavioural patterns, economic strength, technological capabilities, and devotion of public towards the country etc. However, apart from the economic and technological capabilities, these qualities are not easily visible and measurable.

General Schwarzkopf had stated that his army's strength was not only less than that of Iraqi Army, it was grossly inadequate for the assignment, as the attacking force must numerically be three to five times that of the defending forces. Iraqi Navy was so small that it was virtually neutralised in the process of fighting against the economic and trade sanctions on Iraq during 'Desert Shield' phase. The US Naval Forces have two components, one of the US Navy and the other of the US Marines. Marines are unique to the US and have all the three elements viz. army, navy and airforce in them, and these three elements function in a highly integrated manner. The Marine Air Force component in this war was almost half that of the US Air Force component and was equal to the combined air forces of all other nations put together. This difference has not been emphasized in this book, especially because in this war the 'command and control' of all the MNF was totally integrated under one command. The MNF Air Forces were superior to those of Iraqi Air Force, in numbers by a factor of 3 but qualitatively, the superiority was even greater because of their incomparable strength in all aspects of electronic warfare viz. Elint, ECM and ECCM. The SAMs of MNF were also far superior in range, precision and fire power. Although the surface-to-surface missiles like Scud could not match Tomahawks; Iraqi missiles like SA-6, SA-7, SA-9, SA-13, SA-14 and Rollands were as modern and could have defended Kuwaiti airspace admirably, only if they were not crippled. If MNF could gather such formidable Air Forces, what prevented them from improving the Armed Forces ? Was it the fact that they had clearly visualized that firstly, Iraq would abandon the attack and secondly that by using the air power appropriately, they would not need any larger army to liberate Kuwait from the clutches of Iraqi forces.

Strength of the MNF: Air Force

The MNF employed a very strong and well equipped air power, as explained earlier, to achive air superiority over the Kuwaiti air space. The most important aircrafts deployed by the MNF were F-117A, Wild Weasles, AWACS and J-STARS, the details on whom have already been given in earlier chapters. However, there were many more aircrafts which also played a major role for the MNF air force in achieving a decisive grand finale; some of them are described here.

Tornado IDS: Tornado, a multirole, aircraft is an all weather close air support/ battle interdiction, interdiction/counter-air strike, naval strike and reconnaissance aircraft from UK. It has different versions for Royal Air Force, German Air Force, German Navy, and Italian Air Force. It has a maximum speed of 1500 kmph with a ferry range of 3890 km. It has the maximum 360 degrees rapid roll clearance with full lateral control of 4G. It has the facility of secure radio, GPS, IFF radar etc. It is capable of carrying JP-233, AIM-9L, ALARM, Paveway II missiles and 1000-pounder bombs. During late eightees, it was also modified to carry tactical nuclear weapons. Many more modifications were carried out to meet various combat requirements during 1990 for the Gulf-War operations. Though delivered to RAF squadron

Principal U.S. warplanes in action during Operation Desert Storm.

Figure 32

Figure 33
F/A - 18 Hornet

 The High-Tech War of Twentieth Century

Figure 34

USMC Reload Missiles on A F-18A Hornet Aircraft used in the around-the-clock operation to liberate Kuwait during operation desert storm. (Courtesy: U.S.I.S.)

Figure 35
Mid-air refuelling of F-15E by KC-135 Tanker

(Courtesy: U.S.I.S.)

way back in 1982, it was first used in combat only during Gulf-War in 1991.

F/A-18 Hornet: It is a multi mission carrier-borne and land based attack/fighter US aircraft produced by McDonnell Douglas as a low cost, light weight, multi-mission fighter. It was used by US Navy and US Marine Corps as an escort fighter and as an attack aircraft. It has single and two seat versions designated as F/A-18C and F/A-18D respectively, whereas F/-18B is a two seater one. They have a maximum speed greater than 1.8 Mach and their maximum range is 3333km. They can carry upto six AIM-120 AMRAAM missiles (two on fuselage and two on each outboard pylon); upto four Imaging IR Maverick missile (one on each wing pylon). It also has provision for reconnaissance equipment. F/A-18 C/D versions have all weather night attack avionics also.

F-15 Eagles : This aircraft has many versions. F-15 /A/B/C/D are interceptor aircrafts, whereas F-15/E/I/S are role attack / air superiority fighter aircrafts. They have air-to-air as well as air-to-ground mission capabilities. Their maximum speed is 2.5 Mach and maximum range is 4445 km. Its armaments include Maverick, Sparrow, Sidewinder missiles as well as CBU-87 and GBU- 10/12 bombs. Some of the versions (F-15C/D) have digital flight control system with automatic terrain following standard. It also has a high resolution synthetic aperture radar and equipment for flight control, weapon aiming & control, navigation and communications (with cryptograghic capabilities). Its internal ECM system provides automatic jamming of enemy radar signals. Apart from a large variety and quantity of guided and unguided air-to-ground weapons, it can also carry B-57 and B-61 series of five nuclear weapons as well as laser guided bombs.

F-16 Fighting Falcon: It is a single-seat and two-seat multi-role light weight fighter aircraft from Lockheed (General Dynamics). It has all weather and night navigation capability and is suitable for close air support and battle field air interdiction. It has a speed of 2 0 Mach and a maximum range of 3890 km. It has laser range finder, Navstar GPS, LANTIRN capability, anti-jam radio and missile approach warner. It can launch Maverick missile, LGB, AMRAAM (Sidewinder) missile, radar guided Sparrow, sky flash air-to-air missile, IR homing air-to-air missile, HARM and Shrike anti-radation missile etc.

We have compared the strengths of the two sides, mostly, based on hardware and as we have seen that the MNF comes out inferior in ground forces but superior in air power and some hi-tech weapons like Tomahawk, SLAM and PGMS. What is not easily visible is the complex information management system and the weapons of invisible rays. The more the hi-tech weapon systems are expensive and the more concentrated attack is desired, the more is the necessity of a hi-tech information management system, along with its electronic sword and shield. And it is in this invisible domain, the MNF were very strong, and as it turned out later on, Iraq very weak.

Like other inventions, necessity is the mother of high technology. The need, of course, is to win a war. To win a war, an integrated and well knit strategy is to be chalked out. What are the factors on which such planning is based ?

> "Whose army is well equipped with appropriate weapons, whose soldiers' morale is high, that army, without any doubt, would win the war".
>
> 'Shanti Parva', Mahabharat

Chapter 8

War Plans

A complex, dynamic and 'life and death' operation, like war, cannot be undertaken without a well thought out plan. Military thinker Clausewitz has said, "any war plan, however well thought, gets dislocated with the first encounter with the enemy", but a detailed war plan, even then, is a must. Actions of one side and reactions of the other become even more dynamic because even with the best efforts, none of them can get complete information of strengths and plans of the other. Further, quite often, apart from imponderables, the knowledge gathered through various sources could be wrong, as both sides spread disinformation. Therefore planning for a war is a continuous process of monitoring and amending. It also becomes essential that there should be one strategic plan and other tactical plans. The aim of all the tactical plans is to help in fulfilling the strategic plan. Clausewitz had stated, "Tactics constitute theory of use of armed forces in a battle; strategy forms a theory of using battles for the purpose of the war".

A strategic plan depends not only upon the aim of the war, the enemy, the political views, history and the resources of the nation etc., it also depends upon some elements of the neighboring countries, as well as, on the outlook of 'big powers'. The validity of any strategic plan would depend upon all this knowledge, but in any case would remain highly dynamic throughout.

The Iraqi Strategy

Factors Affecting Strategy : When Iraq attacked Kuwait, what were its strategic and tactical plans ? We have had a glimpse of a brief historical background of the war. Despite being a very rich oil producing nation, the eight year long war had imposed a debt of about 80 billion dollars on Iraq and damaged its economic structure. It is possible that, but for its war with Iran, Iraq could have had a surplus of hundreds of billion dollars. The war left the Iraqi Army, the fourth largest in the world, tired on one hand but on the other, well versed and trained in military matters and fully equipped with latest high-tech weapons. Ambitious Saddam Hussein was Iraq's longest reigning president and his dream was to become a leader of the Arab world. He had dreamt that victory over Kuwait would enable him, as the most powerful leader (as far as oil is concered) of the Arab world, to stand against the powerful Israel and resolve the long standing Palestinian issue as per his desire. Kuwait is a small but a big oil rich nation, victory over which would make Iraq one of the richest oil producing nations and also would have given him some maritime rights in the Persian Gulf, another dream and (justified ?) need of Iraq. Probably it was also a testing shot fired by Saddam to assess the international reactions, but he got badly trapped in his own game. He deployed seven armoured divisions and 5,45,000 strong army along the Kuwait-Saudi border in defence. He also started refurbishing the 'air control' concept by redeploying his SAMs and AAAs. He used the tactics of camouflage and dummy targets to deceive the enemy so as to waste his (enemy's) ammunition and time as well as confuse him about the real damage.

The problem must have become serious when on 2 August 90 itself, UN Security Council passed Resolution 660 demanding Iraq to vacate the aggression immediately without any preconditions. This was followed by imposition of economic and trade sanctions on 7 August 90. USA and UK, immediately sent their navies to actually enforce the sanctions. 'What all Saddam had really planned and why', is not known ? One can only make 'a posteriori' conclusions from his actions.

Two Principles of War Abandoned : When, indeed, Iraqi Army made hundred percent defensive preparations to defend Kuwait, it can only be concluded that he had deliberately sacrificed 'mobility' and 'option to attack'. It seems, he also thought that if at all the USA attacks, it would be through Saudi-Kuwait border and not through the Saudi-Iraq border. What made him think so is difficult to guess because he himself had declared Kuwait to be one of the province of Iraq. With these assumptions he did not make any defensive preparations on the Iraq-Saudi border. Strategically this was a wrong decision. Perhaps he also convinced himself that, to liberate Kuwait, attack through Iraqi-Saudi border would not be viable as the prevailing extreme desert conditions would make the logistics support extremely demanding and time consuming for an out-flanking move. In addition, he might have thought that the MNF Army was not large enough to attack such a vast area. With these considerations, he did not keep even a single armoured division for the defence of Iraqi- Saudi border. Saddam might have thought that as his main aim in having a defensive set-up along Kuwait-Saudi border was to convince the US and Saudi Arabia regarding his intention of having no design on Saudi Arabia, it was sufficient to cover Kuwait-Saudi border and more than that would be superfluous. As already stated, Saddam made Kuwait a strong 'fort' with walls made of SAMS and trenches. He might have dreamt that he would be able to enervate and exhaust the MNF, by sitting in his well defended 'fort'. By extending his defensive set-up to a reasonable distance, along Iraq-Saudi border or to cover the Western border of Kuwait would not have gone against his main objective but would have made the out-flanking move little more difficult. This important mistake made the task of Schwarzkopf easier.

Saddam appears to have thus sacrificed two fundamental principles of war -'mobility' and 'offence', which are necessary for victory. By putting his army in trenches behind ditch-cum-bunds, for the purpose of defence, he lost mobility, one of the important principles of war. By making ditch-cum-bund and laying mines along the Kuwait-Saudi border, he made it extremely difficult for his own army to cross over these difficult hurdles, in case he had decided to attack the MNF Army any time. Preparations by Iraqi army would have taken a long time for making an attack, that would be enough to alert the enemy and also that kind of time is not available during a war. Thus Iraq violated the other fundamental principle also i.e. 'offence must be kept as a viable option' during a war.

Kuwait Vis-a-Vis Vietnam : Saddam must have done all this knowingly, but why? One possible reason has already been mentioned at the end of the first chapter, that is, instead of an offensive war against the MNF, which as per his own assessment, he could not have won, a prolonged war was chosen as it might create Vietnam like situation. In those circumstances the USA would be compelled to withdraw its forces, as had happened in Vietnam war. He knew that in a prolonged war, the US would be the worst enemy of the US. He had made strong defensive preparations including the high-tech bunkers and underground constructions to reduce the damages that could be caused by the MNF attacks. He must have known that the USA can sacrifice any thing excepting the lives of American soldiers. He had full confidence on his air defence system for controlling the air space. He must have thought that probability of the MNF Army launching a ground attack was low as it would suffer heavy casualties and he was proven right because the MNF Army attacked only after the 'real bloody war' was over. He might have calculated that if the MNF air forces attack, they will have about 2000 sorties per day and with his strong air defence system (i.e. mainly surface-to-air missiles), if he could impose an attrition rate of 1.2 percent, which is half as much as Vietnamese did, viz. 2.5 percent, he could destroy about 25 aircrafts everyday (US defence analysts had estimated it to be 10 aircrafts per day) and then within a short period of about 15 to 30 days the anti-war lobby would prevail upon USA.

Saddam (and even the USA) could never have imagined that the MNF would be able to establish their air supremacy in such a short period. Hence Iraqis were sure that their army would not suffer heavy casualties in first 15 days. After all, each armoured division of Iraq had 63 modern, quick reacting low level SAMs (15 SA-6, 27 SA-7, 5 SA-8, 16 SA-9), 16 radar controlled modern anti-aircraft quad guns (ZSU-23) and a large number of AAAs. Similarly, every important target in Iraq and in Kuwait was heavily protected with low and medium level missiles SA-3 and SA-2. The SA-2 missile has a range of 50 km and can hit targets up to 90,000 feet altitude. The entire area was thus covered by medium and low level radars to give warning of incoming raid to the SAM units and all the important communications were planned on the 'fail-safe' principle. The command, control and communication centres were safely ensconced in hi-tech underground structures. Saddam had some unconventional weapons also with him which he had threatened to use if cornered. He had already used his chemical weapons in the last war against Iran, and also against Kurds, and nobody had seriously condemned.

Saddam also believed that unity of the MNF would not last long and could be easily broken by provoking Israel. Aggressive Israel would not be able to tolerate death of its innocent civilians. He had modified and upgraded the Scud-B surface-to-surface missiles to increase their range from 300 km to 600 and 900 km only to force such a situation. When launched on Israeli's targets, these giant missiles would kill hundreds of civilias and then Israel would naturally attack Iraq in retaliation. Some Arab nations would then break away from the MNF, internal differences in the MNF would surface and their unity would be damaged. This belief further strengthened his strategy of pure defensive action on one hand and attack by Scuds on the other.

Saddam, with his cent percent defensive postures, made it amply clear that he had no intentions of attacking Saudi Arabia, thereby removing one, and probably the most justifiable reason for the USA to attack Iraq. Further, when its Scuds would kill Saudi Arabians, an average Arab of Saudi Arabia was bound to think that though the war is being fought by USA, only Arabs are being killed on both the sides. This would generate anti-war feelings in them. He would also threaten to put the oil-wells on fire. If even after the threats, the USA would not relent, he would put all the Kuwaiti oil wells on fire and drain out the oil from those wells into the Gulf. Such a danger to environment would alert and alarm the environmentalists and generate anti-war feelings. And finally when the MNF attacks Kuwait, the large number of MNF war casualties and the attendant anti-war feelings would create Vietnam like situation. In addition, during the war many Iraqi civilians would also get killed which would strengthen the global anti-war opinion. And thus by suffering losses, hopefully not too much, if he could prolong the war, anti-war lobby would become strong and by that time it would be summer season with scorching heat and then 'The Vietnam' would not be far off. Arab nations, ultimately would also have a tough leader which they often desired, so Saddam must have dreamt.

The American Strategy

Now, let us see what strategy was planned by General Schwarzkopf. Based upon the past experiences and availability of resources, the US President would have laid down some basic guidelines. He would have desired that Vietnam like situation must be avoided at any cost and hence his first directive would have been, minimum casualties in the MNF, minimum casualties among Israelis and Saudi civilians and minimum casualties in Iraqi civilians, and in that order of preference. The second guideline would have been to conclude the war in the shortest possible period. If they do not succeed in achieving this aim, anti-war feelings would gather momentum in the USA which might create unnecessary but strong hurdles in conducting the war operations. To fulfill these aims, General Schwarzkopf should be given full authority and all the requisite resources he wanted, without any hitch. Further, as per the lessons learnt from Vietnam, any other political decisions should not supersede the strategic decisions and the war should not be stopped until the aims of the Resolution-660 are fully achieved. General Schwarzkopf would be the Commander-in-Chief of the combined multi-national forces with unified command structure.

To fullfil the condition of minimum casualties during the war, it must have been decided to resort to ground attack by army, only at the end. The strategic plan thus was to cause a paralysing attack by air force and cruise and other missiles to destroy the Scud launchers, 'nuclear, biological and chemical' (NBC) weapons' centres and important 'command and control' centres, SAM systems and Air Defence Control Centres. The aims were to destroy information management and decision making centres, communication systems and to establish air superiority by destroying air defence system. This would be followed by establishment of air supremacy with the help of fighter and attack aircrafts as well as cruise missiles to destroy more of the air defence systems, aircraft bunkers, arms, ammunition and fuel depots and airfields etc. Thereafter the remaining command, control, communication systems, the ground based war machinery, vital bridges, roads and depots should be destroyed. When the Iraqi army becomes weaponless, helpless, out of communication, and when its morale would be down in its shoes the MNF armies would be asked to attack and conclude the war. Further, in order to win the war without any political embarrassment, support from as many countries as possible, should be maintained. Further it was essential to contain the war within the shortest period and therefore the attacks must be heavy and carried out relentlessly throughout day and night so as neither to allow time for anti-war hysteria to build up, nor to allow Iraq to repair its damages.

For any Iraqi target close to civilian population, only high precision weapons like laser or electro-optical guided munitions were to be used. Israel and Saudi civilian population would be fully protected. Surface-to-surface missile 'Tomahawk', stand-off land attack missile 'SLAM' and stealth aircraft 'F-117A' were to be used as main weapon systems, initially, to soften the Iraqi air defence system sufficiently. This would enable attack aircrafts, along with their Wild Weasel and other ECM led missions (EF-111, EA-6B and EC-130) and fighter aircrafts (F-15 Sea Eagle, F-3 Tornado and Mirage-2000 etc.) to attack designated targets with the minimum possible attrition rate. In order to keep Israel out of the war, Israel was to be provided with all the protection including the Patriot anti-missile system. The world was to be fed with the progress of the war, regularly and sufficiently.

The Action Plans of USA : Based upon the above principles/policies General Norman Schwarzkopf, the Supreme Commander MNF, and General Horner, the Supreme Commander MNF Air Force jointly formulated the action plans in 4 phases to liberate Kuwait.

Phase-1 : The main aim in this phase was to cause serious damage to the strategic capabilities of Iraq and establish their air superiority within a period of seven days. For the purpose, destruction of 'nuclear, biological and chemical' (NBC) laboratories and factories, Scud launchers, highways, main bridges, equipments, ammunition and oil depots; and impairment of the communications from the rear to the front was to be achieved. Damage and destruction of electricity generation, water works, supply line etc. was to be carried out. To establish the air supremacy, destruction of Army and Air Headquarters, air-fields, runways, underground structures for aircrafts as well as for 'command, control, communication and intelligence' (C^3I) centres, destruction of the headquarters of the four Air Commands, their communication centres and various vital radar and missile systems was to be achieved in this phase.

Phase-2 : Air defence system in the Kuwaiti Theatre of Operation (KTO) was to be paralysed by the MNF Air Force in one day.

Phase-3 : Half the fire power of Iraqi Army in KTO viz. tanks, armed personnel carriers, multiple rocket launchers, anti-aircraft guns and supporting systems were to be destroyed within a period of 3 weeks. Destruction of numerical half of any army, actually reduces the strength by a figure of much more than 2 because the remaining half is left in a totally disordered state specially if they are not given time to recoup.

Phase-4 : Attack on Iraqi Army by the MNF Army with the direct help of the MNF air force to get Kuwait liberated.

General Schwarzkopf had rather abundance of resources. To give examples, they had about 56 AF-117A aircraft (costing $ 120 millions each), 2800 different types of aircrafts, 500 cruise missiles (costing $ 1 million each), hundreds of jammers(costing about $ 0.5 million each) and various space systems which were high-technology products.

Doubts in the MNF Camp : Despite abundance of resources to meet the desired goal there were serious doubts in the MNF camp. Would the absolutely new hi-tech weapon systems which have not been tried in the fields, perform as per their promises ? Would it be possible to establish communication, command, and proper control among the 39 nations' forces with a unity of purpose, despite differences in their training and thinking? Would the Iraqi Scuds, although more of a terror weapon rather than a military one, be surmountable ? Would Iraq use chemical weapons ? Would it be possible to make about 2000-3000 fruitful sorties every day, without the friendly aircrafts getting collided ? Would it be possible to restrain the attrition rate of aircrafts and therefore pilots, very much lower than that of Vietnam war ? Would the weather be too hostile to the war machines and attack- operations, resulting into a prolonged war? And then, would they be able to counter the anti-war sentiments ? It is under such times that the morale of defence forces counts a great deal. Morale is a somewhat intangible phenomenon, neither easy to generate nor easy to be guaged; nevertheless sine qua non to fight for victory. As the MNF had prepared with dedication, diligence and thoughtful planning; they had sufficient sophisticated high-tech resources and the whole world was with them, they were confident of their victory as they believed they were fighting for a just cause. May be it would take slightly longer time than planned and little more casualties than estimated, but win they will.

> "Where there is God of Yoga Shrikrishna and where there is Arjun the best warrior, there shall be the victory."
>
> - Bhagvad Geeta (18.78)

"In the narrative of a military historian, the smallest facts, the most trivial happenings are only the outward signs of an idea which has to be analysed and which often brings to light other ideas, like a palimpsest."
— Proust. (Guermantis Way)

Chapter 9

Kuwait War : The Engagement

16 January, 91 midnight. The preceding day being of new moon, there was that peculiar but known dense blackness scattered all over but this one wore an eerie atmosphere. There was something ominous about this dark night, which apparently did not look different, probably, even to Iraqis. Two missions, consisting of four Apache helicopters each, moved out of a Saudi air base and headed towards the Iraq border. They were armed with night vision goggles which enabled them to pierce that thick inky darkness. Well they could make out the shapes of various structures in the sandy desert, up to 12 km in a limited angle of about 15°. However, all places look alike in a desert and hence inspite of these goggles they were unable to guide themselves properly therefore they had a special helicopter MH-58J to lead their way. This helicopter had a forward looking infra-red system (FLIR), known as, 'Pave Way', which had broader angle of vision than the night vision goggles and was capable of displaying long distance 'pictures' on its scope albeit in synthetic colors. So far, even to Iraqi surveillance radars, if they could see which was unlikely, this would not have seemed unusual, as such missions have been flown for last so many days.

The First Air Attack

After sometime 'Pave Way' signalled that they have reached the pre-designated place near the border and then this guide helicopter returned back to its base. It was not worth the risk for this path finder to go any further. Their target was not far off from this place and they were confident, at least they wanted to feel so, that they would be able to cover this distance with the help of their night vision goggles. Mind you, all that those pilot could have done was to have thoroughly studied the map of that area and had located 2-3 check points on the ground. The pilots were still apprehensive, although Iraqis may not be able to see them, as they had planned a safer route based on the knowledge of Iraqi 'electronic order of battle'. The early warning radar which they were going to attack were the first set of radars across the border and therefore these targets were extremely important. Some Iraqi soldiers might see them with the help of night vision devices, soldiers who might come that way merely by chance because they had also found out, in advance, that no Iraqi soldier was deployed on the route.

Flat Face Smashed

They were afraid, but not enough to loose heart, perhaps enough to give them courage for marching ahead. They were not afraid for their lives but for their mission. They knew that their missions are very important because only after they destroy these radars, a large number of attack missions would be able to fly through the short safe passage thus created into Iraq. One of these early warning radars was 'Spoon Rest', antenna of which they should be able to recognise at the first sight itself as, it looked like a spoon rest, a spoon hanger. The other was 'Flat Face', an old technology radar, operating at 200-300 MHz e.m. frequency. Developed countries had stopped using them long ago and hence the US Air Force had produced very few jammers suitable for it. In view of this also it was essential to destroy them first. Another radar was 'Squint Eye' antenna of which was similar in looks to that of 'Flat Face' - like a face with flat nose. This radar, in addition to early warning, also provided low level acquisition for S-3 missile guidance radar, but they were not to go anywhere near the S-3. To be extra safe, unlike their standard practice, they were flying at very low level, though as a daily routine, for last few hours the 'Prowler' (EA- 6B) and 'Raven' (EF-111) were radiating jamming signals. They knew that 'Prowler' and 'Raven' had very effective jammers and in many exercises they have never let them down, but this nagging feeling about the jammers worried them as on one hand, these work with invisible rays and on the other, they can not see the effect of jammers. What, just in case some of the victim radars may not be getting jammed or fully jammed ! Then an Iraqi night fighter aircraft would be able to shoot them like a sitting duck. Such thoughts were bothering them but not really worrying them.

Inauguration by Hellfire

Even in that ocean of darkness they could see the spoon hanger antenna from a distance of 12 km. But they did not want to take any chances and hence went closer to it for confirmation. The second mission was also able to locate its victim, and turned towards it. While confirming the identity they came upto 7 km from the radar, they thought another 1-2 km and they would let lose hell by launching the laser guided missile - 'Hellfire'. They looked at their watch, it was 0238 hrs, well before the dawn of 17 January 91. They were 8 minutes behind schedule; of course they got delayed in reaching the border itself, perhaps because of the headwinds instead of crosswinds. In future they would have to do better planning, they thought. As soon as they reached 5-6 km, they launched a salvo of Hellfire missiles with a gap of 10 seconds between each missile. Immediately they aimed their laser designator and illuminated the antenna of the radar station for 10 seconds, then the transmitter for 10 seconds and then the electric generator. By then they were just 3 km away and the Iraqi AAAs started shelling left, right and centre. Since they did not require to go further, they turned towards the other radar, fired on that also successfully and turned back. The other mission was also on its way back. They knew that it was also successful, not only because they heard the bang but mainly because, the pilot of the other mission could not control himself despite all the instructions, and while launching last of his Hellfire, he shouted on the radio: " This one is for you Saddam". They were overwhelmed with their grand success. Though very keen to share their joy, they had to use all their will power to maintain silence and they kept mum till they crossed the enemy territory. Overjoyed they should have been, as they had destroyed the special difficult-to-jam radars and created a safe passage for other attack missions. Thus their weapons inaugurated the Air-Operation of'Desert Storm' by blowing a bugle with a hellish bang. One Tomahawk had already attacked an important target in Baghdad, as if to pip the ribbon of the first attack on Iraq. 52 Tomahawks were fired in the initial attack.

Entering Stealthily

Chuck, the pilot of F-117A, looked at his watch, it was wee hours of 17 January, 0130 AM; "well A.M. does not sound as midnight", he thought, "I had taken-off from the Tonopah airfield at 0040 hrs, forty minutes past midnight. Habits could be funny. Actually there is a place called Tonopah in USA and there is a place in desert called Khamis Mushait, but all the pilots called 'Khamis' as Tonopah - well, Eastern Tonopah. Khamis is situated at 6800 feet above mean sea level whereas Tonopah is at 5500 feet - difference of only 1300 feet. At least the place is.cooler than other parts of the desert, though very far off from the border, about 1200 km in the South-West corner of Saudi." The thought came to him automatically, "After another 10-15 minutes of flight I have to get refuelling done before reaching the border. At least something to do on such a long, lonely and deserted route. Most of the other aircrafts have many things to be attended to but this one needs nothing on the way excepting keeping an eye on the instruments and as per the instructions of Jack (the four dimensional navigation system of F-117A nicknamed as 'Jack') to change the direction or speed. Excepting dry sand, nothing is seen during day time in the desert, and during nights, not even that. I wonder, if this shadowless sooty ocean of darkness is preferable or the burning, shimmering wavy ocean of heat during the day time". Well this darkness gave him a feeling of safety, his aircraft became one with the jet black night.

"Back home, night flights were never as dark as diving in a bottomless deep and dark sea. There I could see twinkling stars below as in the sky above. How must Jane (his wife) be spending her time, it's ten days since I received her last letter. She promised to write every week and in fact, wrote regularly. I know Jane loves me, loves me so much. Poor thing, must be suffering now. But Jane is a brave girl. Obviously, the transport aircrafts and courier aircrafts would be giving priority to -weapons and supplies. Anyway the letter would come tomorrow if not today". When he looked left, he saw his friend, Dick, flying by the side of him. Today he is the first pilot and Dick is following whereas in the next sortie, Dick would be the first pilot and he would follow. Now, both have the same target. Dick is expected to fire only if Chuck's precision laser guided bomb, due to some unforeseen reasons, does not hit the target accurately, but today to be doubly sure to destroy the target fully, Dick would launch his bomb also. He did not want to take any chance as the target is very important and being the first target of F-117, the success of other attack missions would depend upon its destruction. An important communication centre in Baghdad city is to be destroyed. Later on, other aircrafts would attack the nuclear factory etc. located further ahead.

He heard Dick saying, "Chuck, Left, 11'o clock about 20 km. away." Chuck saw the tanker KC-10 circling at the designated point. By now 'Jack' also gave a 2 minute warning of the tanker. Both of them got the gas filled in the air itself and started their ascent to move towards the border. At 30,000 feet, only the radar -tracking and guidance - of SA-2 has the capability to track high altitude targets but even those would not be able to see them. They have another 20 minutes to reach the border which they would cross at 0220 hrs, whereas the scheduled time for the target was 0250 hrs.

Last evening he had woken up at 5 p.m. and by 8 p.m. he reached the operations room for mission briefing and other preparations. He planned the complete flight details with the help of a mission programming computer system in just 15 minutes; and wasn't he glad, earlier it used to take 3 hours. All check points, where, what speed, which direction, what altitude, distance of the next check point, where to start descending before reaching the target, where to turn and how much, what to look for, how to identify the target, when exactly to lock-on to the target, when to press the button for releasing the bomb and all that were planned and calculated. There might be a delay of a minute or two in reaching the target as the winds in Iraq are not predictable accurately. 'Jack' gave a warning that the check point on the border had arrived. They were just one minute behind schedule, probably the refuelling took that extra minute or may be the winds, Iraqi winds have caused that delay. They moved towards right from the check point. Jack asked him to increase the speed from 450 knots (nautical miles per hour. 1 n.mi. = 1.8 km) to 465 knots, and change the direction by 1° towards port (left). He looked at the watch; it was 0221 hrs. He thought that now he should be alert, after all he was entering into the enemy's airspace, it was not an exercise or even a deception flight like last night.

At 0245 hrs the last check point arrived and now after 2 minutes they had to steer the aircraft 90° starboard (towards right) and descend to low level for releasing the bomb; they had to drop the bomb at 0250 hrs. 0249 hrs, now see the target, in FLIR image display, in wide angle, right there is the target, 0249 hrs 30 seconds, now look in narrow angle, the target is clear, now look in the display for the circle, OK, it is on the target, press the button. The target was now under the aircraft, now the bottom display (DLIR) would be tracking it, the laser beam must be illuminating it, and bang !!!! And 3 seconds later, another bang !! Dick had also released his bomb. Simultaneously the Iraqi guns started firing randomly as if somebody had woken them up from their slumber suddenly. Start ascending in the sky and get back. After landing, look all of it in the video film. How the IR imaging tracked the target and how the bomb was directed towards the desired aim with a high level of precision and fineness. The next assignment for another F-117A was to drop a bomb on an army communication centre located in the Baghdad city, exactly 0300 hrs, on the dot.

Massive Air Attack

Throughout the night on 16 January 91, F-117As went on bombing Iraq's medium and low level missile units, air defence radars, communication centers, Scud launchers and 'nuclear, biological and chemical' weapon factories. In the morning, at 0330 hrs, hundreds of, 700 to be exact, F-15E, F-16C, A-10A, F-111, Tornado GR-1, Tornado GR-1A, F/A-18, A-6E, A-7E and Jaguars for attacking the targets, and F-15C and Tornado F-3, F-14 and Mirage-2000 for fighter escort and establishing air superiority, AWACS E-3A and E-2C Hawk Eye for aircraft warning and control, special aircrafts for air traffic control, and EA-6B, EF-111 and EC-130 for jamming radars and communication centers, F-4G for destroying missile radars, KC-135 and KC-10 for refuelling, took- off in suitable formations and penetrated through the strategically planned and chosen routes to carry out their attack missions for important Iraqi targets.

Important Targets

The important targets of Iraq were air defence (early warning and acquisition) radars, missile control radars, GCI radars, air defence communication centers, Scud launchers, 'command, control, communication and intelligence' (C^3I) centres, air defence direction centers viz. Kirkuk, Rutba, Nasiria and Baghdad, Army and Air Force Headquarters, 'nuclear, biological and chemical weapon' factories etc. Amongst them, most critical ones were being destroyed from pre-dawn to early morning by F- 117A and Tomahawk missiles. Aim was to damage the brain (information and command centre), the eyes (radars) and ears (communication centres) of the Iraqi war demon.

Iraqi Air Force Could Not Fight

Though the multi-national forces were aware that most of the Iraqi air defence network has been paralysed in the night itself, they were astonished when they met with unexpectedly low resistance from the Iraqi fighter aircrafts. Iraqi Air Force was making about 100 sorties per day during Operation Desert Shield, but on the first day they tried to make only 30-40 sorties, that too only during day time. Only 3 MNF aircrafts were shot down during first 12 hours, one by the enemy's SAM and the other two by AAAs. Loss of only three aircrafts in 1000 sorties in the enemy territory was an achievement beyond expectations. The first day, 8 Iraqi aircrafts were damaged in the air combat and the second and third day 9 aircrafts each. During the first three days the Iraqi fighter aircrafts made only 40-50 sorties each, much less on other days and before the end of the week the Iraqi aircrafts more or less gave up fighting.

The day Iraqi aircrafts took part in the air combat they, on an average, lost 5 aircrafts. No air force in the world can afford to lose 5 aircrafts in just 30-40 sorties, nor can it put up any worthwhile fight. With an attrition rate of 10-15%, any air force would be grounded in a few weeks time. It means that Iraqi pilots wanted to fight, they even accepted a heavy attrition rate, but when they just could not really fight, they had to give up. They flew only during day time, also confirms that they could not depend upon their air-defence system. Their excellent aircrafts like MIG-29 and Mirage F-1EQ-6 could not face those of the MNF, which only means that radars and communication systems of Iraqi aircrafts and GCI system were either fully jammed or destroyed and at the same time the radars and communication networks of MNF were working efficiently. With an electronic supremacy like this air superiority and then air supremacy is bound to be established.

Time for a Bold and New Decision: When the Iraqi aircrafts did not remain safe even in the reinforced concrete bunkers, they decided to send them to Iran. This action further conforms the defensive strategy of Iraq. In air combat, the Iraqi Air Force was continuously losing, and their air defence systems as well as other important targets were getting destroyed at an alarming rate. Such a critical juncture was the time for taking bold and meaningful decisions and not an irrelevant one, like sending the aircrafts to Iran. Taking such major and bold decisions, was it beyond the capabilities of a dictator ?. The whole strategy as perceived by Saddam was already dead, and only a new and more mature planning could have breathed a fresh life into the MNF- Iraq war or Iraq. Saddam should have either taken offensive with his army or accepted the UN resolution 660. Obviously, having abandoned offence in his original strategy, the second course would have been more pragmatic.

Heavy Attack Schwarzkopf and Horner were faithfully sticking to their principle of 'heavy attacks'. Despite the success beyond expectations, they did not relent. Apart from making 2000 sorties per day for the first three days, 196 Tomahawk missiles costing $200 millions (!!) were launched in those three days. Later the requirement of Tomahawk got reduced, a credit to the MNF strategy, and only a total of 300 of them ($360 millions) were used during the complete war.

Low Versus High British Tornado, Buccanear and French Jaguar attack aircrafts, as per their practice, were flying low due to the fear from enemy's strong air defence system whereas the US aircrafts, as per their principles of air safety based on heavy jamming, were flying at high levels. In air defence system, if the enemy is using missiles as well as interceptor aircrafts in a balanced manner, it is always preferable to sneak into its territory by flying low. In case he depends mainly upon surface-to-air missiles, flying at high altitudes is better. However, in both the cases jamming of radars and communication centres is absolutely essential. While in low flying, the danger from AAAs (unguided) and other small arms like guns, stenguns increases whereas the danger from interceptor aircrafts is less.

In the first four days, four Tornadoes were shot down whereas the casualty figure for US aircrafts was rather negligible in comparison to their number of sorties. From fifth day onwards the British as well as French aircrafts also started flying at high altitude, of course, by that time MNF air supremacy was also established.

High Intensity and High Tech High-tech weapons were being used in abundance. The aircrafts A-10, F-16 and F-111 etc. were launching, on an average, 100 Maverick air-to-surface IR guided missiles per day. Maverick missile is not a very costly munition, at a cost of $100,000 each, it is not cheap either. Similarly Hellfire was also used liberally.

Surprise With Noise Jamming !

Electronic warfare was being practiced well before the night of 16 January 91. Though the radar gets blinded by the electronic noise jamming, it also announces the arrival of attacking forces and gives the enemy a chance to start trying some counter measures or other. That means, advantage of surprise is lost from noise jamming to some extent. However, to get this advantage of surprise, the enemy is tactically deceived by resorting to noise jamming even when no attack is being planned. As per this doctrine the MNF started the process of jamming 3-4 days before 16 January 91, and simultaneously, every night, the attack aircrafts, in formations, used to fly from Saudi to Iraq border and return, as if they were carrying out war exercises. But on the night of 16 January (morning of 17 January) at 0130 hrs., attack aircrafts came in formations and also returned, whereas different sorties of the stealth aircraft F-117A and Apache, instead, penetrated into the enemy territory for attack.

Dynamics of Jamming The electronic jamming is dynamic in nature. The MNF saw that the AAAs of Iraqi air defence system start firing as soon as they start jamming and stop for some time after 4-5 minutes, probably to give time to the gun barrels to cool down a little. Hence they used to start jamming with full blast, 5-6 minutes in advance, and when the AAAs used to slow down after 5 minutes of firing, the attack aircrafts used to complete their missions. But such a tactics cannot become a regular pattern.

Dynamics of MNF Plans

Though it was still the First Phase of attack as per the war plan, Schwarzkopf simultaneously started the second and third phase after the air superiority was established in 3-4 days. After gaining an experience from destroying the Iraqi air defence system, he easily destroyed, in the first phase, the Iraqi air defence system of Kuwait also, a task reserved for the second phase.

Mobility is Essential in a War

It took only three days to destroy the static locations of Scud missile but their mobile locations gave a tough time to the MNF till the last day. Schwarzkopf deployed all the available resources like satellite, surveillance aircrafts, J STARS, E-8 attack aircrafts, air superiority aircrafts etc. to search and destroy the mobile Scud launchers. Though the MNF had established their full control over Iraqi skies, Saddam's mobile Scud launchers could still fire 2-3 Scuds on the Israeli and Saudi civilians every mid-night. These mobile Scud launchers proved to be an excellent example of the principle that mobility is very important in a war.

Precision Versus Dumb

During the initial days of war, F/A-18 aircrafts, were given the responsibility of destroying all the bridges on the Iraq- Kuwait supply route. F/A-18 is a multi purpose aircraft and is one of the best in the world. But during this war, time was of essence, due to still surviving air defence/SAMs and use of conventional weapons, they could achieve only limited success in day attacks, hence this task was later assigned to F-117A. These aircrafts then destroyed 40 out of 52 bridges by the third phase.

A Mission of 60 Attack Aircrafts Unsuccessful !

During the preliminary days F-16 aircraft was assigned a mission to destroy an important nuclear weapons factory situated in the Eastern part of Baghdad. All possible help was provided to ensure the success of this mission, which comprised 32 F-16 aircraft assisted by 16 fighter escort F-15C, 4 EF-111 - the best electronic warfare escorts, and 8 F-4G, the radar killers

(Wild Weasel). Since the target was deep inside Iraq, 15 KC-135 tanker aircraft were also pressed into service for refuelling, midway in the air, all the 60 aircrafts of the attack mission. The mission was sufficiently big in conformity with the principle that the attack should be heavy, only then the success could be ensured. En route, the mission jammed all the radars, destroyed the important air-to-ground missile guidance radars and reached the designated target by penetrating the strongest possible air defence system of the enemy. On reaching the target they found it fully enveloped in a cloud of smoke, which was spread by Iraqis as a safety measure against such attacks. The complete fleet of 60 aircrafts came back disappointed. They could have done blind carpet bombing, but that was undesirable. Creation of a smoke screen is so well known a tactics that it has given birth to the famous proverb on its name. Even then, the MNF did not anticipate this before launching, thereby resulting in, probably the most expensive and unsuccessful attack mission in the history so far.

10 Aircrafts Successful

The same night this challenging task was assigned to 8 F-117A aircrafts. They hardly needed two small tankers, KC-10, and two EA-6 (Prowler) jammer aircrafts just to be sure ! Their forward looking and downward looking IR systems could see even through the smoke screen and they could successfully complete their mission by dropping their laser guided bombs precisely on the targets.

Thus the comparative performance efficiency of stealth aircrafts vis-a-vis visible aircrafts as well as precision guided weapons vis-a-vis ordinary weapons was demonstrated and this conclusively proves the supremacy of high technology systems in a war.

Mobile Scuds Delay Third Phase

In the third phase, the MNF wanted to reduce the strength of Iraqi Army deployed in Kuwait to half, destroy all their bridges, damage all the roads leading to 'Doab' (Mesopotamia) at the key places and cause heavy damage to storage depots, power houses as well as the water supply system. They were carrying out these tasks but a major part of their air force was diverted on searching Iraqi mobile Scud units and destroy them. In addition, many of their missions became victims of the prevailing bad weather, their third phase could be concluded in 38 days, 8 days behind schedule, despite an early air supremacy.

Knowledge of Battle Damage Important

Assessment of battle damages is very important and a difficult task, the truth of which was clearly visible in this war. If the damages are wrongly assessed or not assessed at all, the aircrafts will be sent again either to determine the damages and if necessary attack or attack the already damaged targets without determining. Alternatively some targets will be skipped if wrongly considered to be damaged or destroyed. Though MNF had the best resources viz. satellite, Tornado GR-1A, Jaguar, RF-4C, U-2R, Mirage F-1 and J STARS E-8, even then it seems, they underestimated the battle destructions; or was it a deliberate ploy when in fact the MNF wanted to destroy, may be 70% by air power and not 50% as slated. Wherever the MNF Army entered Kuwait, the fire power of Iraqi Army was always found to be far less than expected due to which the progress they achieved in Kuwait was always faster than planned.

Sitting Ducks It is also possible that to play safe, the MNF Generals might have decided to cause more massive destruction of Iraqi defence system than really necessary, as they could do so with ease and safety within the original stipulated period. Further, it made the task of their army that much easier and reduced the hazards in ground battles and consequently reduced their casualties during the attack. After all, A-10, Apache, Cobra and Black Hawk aircrafts, by their laser guided bombs and missiles, were destroying the Iraqi tanks, guns and other targets like sitting ducks.

Old Bomber New Fire

Two A-10 aircraft destroyed 21 tanks in two sorties. All the

Figure 36

Black Hawk: The US Second Generation Technology Transport Helicopter

three types of helicopters jointly destroyed about 200 tanks per day continuously for 6-7 days. The giant aircraft B-52G, old but reinvigorated with modern navigation and weapon control system, dropped about 72,000 bombs, weighing 25,700 tons, in Kuwait and Iraq. This was about 31% of the total bombs dropped by the US Air Force. Once the air supremacy was established by the MNF, the war operations got simplified, not only for the air force but for the army too. Without air supremacy neither the reinvigorated B-52G nor the attack helicopters could have performed the magic, so easily and with impunity.

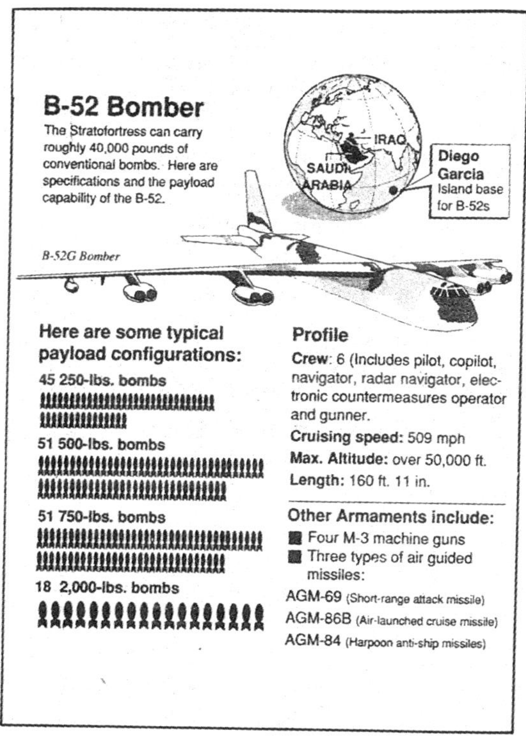

Figure 37
B-52 Bomber

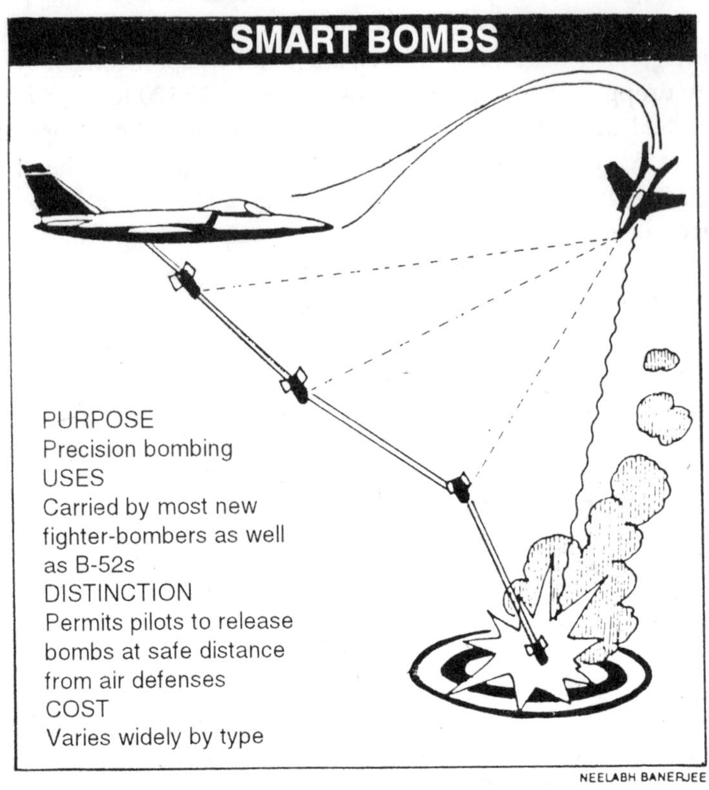

Figure 38
Smart Bombs

Operation 'Desert Sabre' : Sabres Didn't Have to Rattle

In the fourth phase i.e. 'Desert Sabre', the MNF Army attacked Kuwait but still with the help of air force. Schwarzkopf took tremendous advantage of 'surprise' and 'deception' techniques. The mock attack on the eastern coast of Kuwait by US Marines was only to deceive Iraq, whereas he planned the real attack, not so much from central but from the west direction. He first launched an attack from the centre, as Iraqis had expected, to strengthen their (Iraqis) confidence on their strategies, wherein they did not expect an attack from the western side. Then, he planned to enter Kuwait from the western side, out flanking the Iraqi Army, quickly reach north of Kuwait, and suddenly attack the Republican Guards - Iraq's famous armoured division - between Basra and Nasiria. This outflanking move could be made at lightning speed in that desert because of the air support, this time including major transport operations. The 'blindness' of Iraqi forces further helped in maintaining the 'surprise'. It was done so fast that the Republican Guards were caught on the wrong foot, even for their retreat. In this battle, the Main Battle Tanks (MBT) M-1A-1 of the US destroyed the rest of the 200 Iraqi tanks whereas they (MNF) lost only one, a fact demonstrating that the so called most ferocious tank battle of the Kuwait war was extremely one sided. Indeed, the Republican Guards were caught retreating with their will to fight totally crushed by the MNF air power.

Another fierce battle was expected on Kuwait-Saudi border where Iraq had erected strong defences (as described earlier). The Seventh Corps of the USA and the First Armoured Division of Britain were assigned the task of attacking through the western border between Kuwait and Iraq. The First Armoured Division was responsible for the tri-junction area at the border between Kuwait, Iraq and Saudi and was to enter Kuwait through the southern part of the western border between Kuwait and Saudi Arabia. The Seventh Corps was responsible for the area adjacent to that of the First Armoured Division and was to enter Kuwait from the northern side of the western border between Kuwait and Saudi Arabia. The Seventh Corps was to enter Iraq's boundaries first and the 'G-day' (i.e. go day) for the army was 0200 hrs, 24 February 91.

Saddam's Dream of Trench Fire Extinguished

A day before the G-day, the US attack aircraft A-10, with the help of electro-optically guided Maverick missiles, destroyed the valve, through which Saddam Hussain wanted to fill the trenches, dug along the Kuwait-Saudi border, with oil for putting it on fire. Whatever oil was left in the trenches was also burnt. The mines which were laid in the barbed wire jungles, beyond the trenches, were also destroyed, as far as possible, by anti-mine weapons launched from aircrafts. The army was moving very cautiously in the mined area after thorough investigation, and wherever they found a mine they cursed the airforce for not having done a good job. It also indicated how much the MNF' Armies wanted to depend on their Air Force.

Robert Fox Reports

Robert Fox of Daily Telegraph accompanied the First Armoured Division to the tri-junction battlefield to report an eyewitness account of the war. Apart from others, this Division contained Fourth Armoured Brigade, part of which was 14/20 King's Hussars, with which Robert Fox was attached. He published his eye witness account '100 Hours of War' in the 20 April 91 issue of 'Daily Telegraph', certain excerpts from which are given here to illustrate the complexities and realities of the land battles.

Mother of All the Traffic Jams "Iraqi Army had constructed bunds along the trenches in such a way that when tanks would cross over them, their belly would be visible to the Iraqi soldiers who could then take advantage of firing on the soft belly of the otherwise armoured and strong tanks to destroy it. This tactics was a topic of discussions all over. When Fox's vehicle reached that point, his reactions of the event in his (Robert Fox) own words were, "with lots of efforts we had spent hours in practicing to cross over the bunds, but in actuality it was the other way." He added further, "after successfully climbing the bunds, stores supply convoys were found in such a large number that Saddam's prophecy that "this war will be the mother of battles" became the 'mother of traffic jams'.

Army Discovers A Tank's Graveyard "While marching ahead, Col. Vickery's tank destroyed an enemy's T-55 tank and when he got down to see the extent of damage in that area, they found hundreds of totally damaged tanks and dead Iraqi soldiers without any identification marks. They were either completely burnt during the air attack or got submerged into the sand, whereas the wounded ones must have fled away."

Tank Batteries for Bunker Lights "The tank which was destroyed by Col. Vickery was in the outer part of a large (49 sq.km) Iraqi protected area. As per the Commander Fourth Brigade, this Command area, containing underground bunkers, defensive systems and weapons stores depot etc. was so large that it could comfortably accommodate one Iraqi Division. The underground bunkers were so made that the Iraqi soldiers had all homely comforts in them. They had removed the batteries from tanks and other military vehicles and fitted them in their bunkers for getting heat and light. Hundreds of electric generators, water tanks and bicycles etc. were scattered all over. Though that place was strategically very important, the complete atmosphere was engulfed in fear and disappointment. Tanks were positioned in so deep trenches that it was not only impossible for them to come out, it was difficult for them even to turn their barrels. On the whole, the Iraqi commanders did not have any idea of the importance of mobility in a war. On 25 February 91, G+1 day, at 2250 hrs the Challenger Tank started shelling on a new enemy's position. They saw four tanks and some other vehicles running away under the umbrella of rain. Major Vicks, Commander D- Squadron, who was protecting our War Group from the rear as well as sideways, reported that, " 2 tank squadrons of Hussars have gone ahead but the Iraqi soldiers are still coming out of the underground bunkers and are surrendering."

Tank March : A Glorified Taxi Service "Major Gilespy fired the 'light emitting shells' to see the enemy's condition in dark, suddenly they saw Iraqi soldiers hoisting white flags for surrender. The guards of the 'Queen's Company of the Grenadiers' were instructed to move ahead with their Warrior vehicles and takeover the charge of the prisoners. Major Backer, the Company- in-charge, in a protesting tone, said, "I feel I am running a glorified taxi service." They came to fight a battle with the Iraqi soldiers in the underground bunkers with guns, revolvers and hand grenades but here the Iraqi soldiers were impatiently waiting for an opportunity to surrender."

Ground Attack Manoeuvre "We used to doze off in the Warrior vehicle. Whenever we stopped at a point even for 2 minutes, the driver also slept. As soon as the enemy started shelling, the Battery Commander woke him up with a kick. We prepared and drank coffee in that cosy vehicle and somehow spent the whole night with difficulty. Early in the morning we saw on the horizon, the flames of burning tanks trying to swallow the darkness. Suddenly on their own, the forward tank squadrons started shelling. Now a very efficiently executed ground attack manoeuvre was witnessed. The attack on 'Target Brass' (pseudo name of one battle field) required two hours of fast run through the enemy's tank companies and other combat vehicles and shell anything found moving. Half way through their attack, they suddenly turned at an angle of 90° - a very difficult manoeuvre - very expertly executed - so much so that a Sergeant on the parade ground could also take pride in such a capability and fitness. Suddenly, we received orders to stop as the 'Target Brass' had been won and there was no casualty on our side. On 26 February 91, G+2 day, by afternoon the Fourth Armoured Brigade had completely finished two brigades of Iraqi tactical reserve."

Battle : A Stroll in the Garden: " After we got down, we saw prisoners and prisoners all over. The gunner of our vehicle said, " this turned out to be like a practice exercise excepting that there was an enemy in between - like taking a stroll in the garden." Brigade Commander Brigadier Kardingli said, " we were told that there are 5 Iraqi Divisions reserved in this area and we have to attack two of them, 12th and 52nd. I do not understand what resistance did we get on the way, during attack."

In This War Do Not Fire At The Enemy: " The next mission was to attack 'Target Copper' which contained a major communication centre. The Royal Scotts Dragon Guards War Group attacked that communication centre. The Commander, Lt. Col. Sharples said, " we saw a long climb from the rear of which the Iraqis were firing machine guns. When our soldiers retaliated with a return fire, the Iraqi soldiers tried to run away or hide behind their combat vehicles. Whenever they fired, we returned the fire. When we moved Eastward, we destroyed a lot of tanks and vehicles. The infantry positions were clearly visible with the help of light emanated by the fire weapons launched by the tanks. The C-Squadron also destroyed everything it found moving on the way. Then we stopped to ensure that we do not attack each other in the darkness. The next target was a few tents near a well, located at some distance. We could see a glimpse of white flag flying even there. Thinking that this might be a hospital, we waited upto the dawn. And with the first ray of sunlight, the Iraqis started surrendering and that too in hundreds. After this Brig. Kardingly issued instructions that now in the battlefield no body should fire at the enemy, instead they should fire only warning shots above their heads."

Hunting A Jungle Fowl: "Similarly after the attack on 'Target Platinum' was over, Col. Rogers, who took 300 prisoners of war (POWs), said, "This attack was like hunting a jungle fowl in the forest". Instructions were issued to win 'Target Lead' and reach the phase of 'Smash Border' and it was estimated that this operation would take 10 days but the task was finished in just two days. In absence of any intelligence, the Staffurdus, a regiment of Seventh Armoured Division, found themselves face to face with a large and secured enemy area. The area was being defended by a battalion and the attack was also of the same strength."

The only Casualty in A Regiment : "The forward company fired warning shots on the first row of trenches and the enemy became jittery. Col. Rogers fired 6 shells which had hit 6 bulldozers (as confirmed later). Warrior combat vehicle carefully moved ahead - the two companies of the first row of trenches surrendered. As soon as soldiers of the Staffurdus Regiment got down from behind, to arrest their prisoners, the Iraqi soldiers started machine gun fires. An Iraqi version of deception tacties which the rest of the world would like to identify as treachery and is recognised as a crime in the international law. The only casualty from Staffurdus was a soldier named Stephen Molt."

Attack on a Target Without the Enemy : "During the same period, the Queen's Dragon Guards had to struggle for 6 hours on 'Target Lead' and it was the longest British mission. However, during night, when all the three war groups of the Seventh Brigade surrounded 'Target Lead', they did not find any trace of the enemy." !

Blue on Blue : "On that night the Fourth Brigade headquarters was completely grief stricken. That day when Fusiliars, a regiment of the Fourth Brigade, were engaged in a gun battle with the enemy, under the 'Target Steel' mission, their two Warrior vehicles suffered a blast. They were hit by two Maverick missiles fired by (surely by mistake) the US aircraft A-10, from a height of 1500 feet. Nine Fusiliars were killed and eleven injured and, ironically, this was the biggest British casualty during this war. This was an example of 'Blue on Blue'[1] in this war. Due to this unfortunate accident, the attack on 'Target Steel' was temporarily halted for few minutes, and then the Target Steel was captured in an hour's time.

Madina and Hamraki : "On the afternoon of G+2 day, Brig. Hammerback was informed that as per an intercepted message, the armoured division 'Madina' of Iraqi Republican Guards had been instructed to come out at 1800 hrs, the same evening. Madina and Hamraki were two best Iraqi Divisions, which served as war reserves for Iraqi Armed Forces. They were Saddam's favourite and contained the highest paid soldiers as well as the best tactical systems. Both the Divisions were to provide protection to Iraqi Army returning from Kuwait to Basra. The MNF soldiers were eagerly waiting for this moment. As soon as the enemy tanks came out of the trenches, the MNF tanks and anti-tank aircrafts started shelling them. ... .."

Enough Resources to Prolong the War : "When we went to see the place which we won after 10 minutes of heavy bombardment, we were stunned. The Command Post was

---

1. Friend killed by a friend. The friendly forces are always shown by blue colour in the battle field maps or war models and hence the proverb.

decorated and laid out with costly chairs and sofa sets. The operator's room had grid maps and other complete details about the defensive artillery range areas towards south and east, the directions from which (unfortunately) there were no attacks. They had the best electronic intelligence (ELINT) systems. One of the Iraqi intelligence map had depicted the hour by hour progress of the Fourth Brigade. Dummy artillery guns and aluminum reflectors were sensibly put here and there to deceive the attack aircraft pilots and their radars. On the whole Iraq had enough resources to prolong the war and make it tough for the MNF, but their morale and will to fight was crushed by air attacks. Britain lost only 17 soldiers in this battle; and 95% of the 'Challenger tanks' and 'Warrior' armoured combat vehicles were functioning perfectly well till the end of the war."

Scorched Earth : Environment as a Weapon

When an attacking enemy is so powerful that discretion becomes better part of valour, 'retreat' is also a valid tactics of war and has to be planned carefully. Depending upon the strength, ambition and aim of the attacking army, scorched earth policy, as was carried out by Russians in response to the attack by Napoleon on Russia, is sometimes followed. The objective of destroying one's own valuable resources is to make the task of advancement of the enemy difficult, if not impossible. Ultimately, it was Napoleon who had to beat a hasty 'retreat' from the Russian soil, mainly due to this scorched earth policy and the treacherous weather.

In case of Kuwait, scorched earth policy cannot be justified, but Saddam executed it in a demonic way. Indeed, he used environment as a weapon, probably for the second time in the history. The first use of environment as a weapon, was again made by Saddam in 1984, against Iran, when he burnt oil wells at Abadan. The US had used defoliating agents (inhuman, in any case) in Vietnam war in which environment was badly damaged. However, the aim then was not to use it as a weapon, but to clear the forest so that the enemy's movements could be seen. Atomic bombs were used in WW-II to destroy the two cities' population, buildings et al. Saddam put all the 600 oil wells of Kuwait on fire while retreating. This burnt almost 1000 million tons of oil[1] in a period of about 9 months and it caused enough smoke to cover about 15,000 sq. km. of area[2]. It also caused acid rains in Turkey, Iraq, Kuwait and Iran etc. The spilled oil reached the Gulf and caused death of millions of sea creatures and birds, a holocaust of demonic proportions. The deep soil water resources and the soil as well were also polluted by the oil which had seeped in. In the affected area, the smoke has increased the respiratory diseases to an alarming level, mostly affecting children and older people. All these human and environment losses have been, fortunately, lesser than feared earlier but were tragic, inhuman and meaningless. All this loss of environment and human sufferings can not be measured in terms of dollars and there is need to make appropriate international laws and to find means of enforcing them. When environment is attacked, nobody is a gainer but the whole family of mankind is a sad loser.

"My argument is that War makes rattling good history; but peace is poor reading." - Thomas Hardy : 'The Dynasts'.

---

1.   This amount of oil could have satisfied the demand for more than 10 years for a country as big as India.

2.   However, Richard Turco of University of California had earlier reported in 'New Scientist' that such a large scale burning oil could release three million tons of black smoke which may shade up to 100 million sq. Km. of surface.

—Anonymous

Chapter 10

Lessons From the War : An Analysis

Iraq has too many wounds to nurse after the war. There is no way by which it can balance its losses in a short or even a medium period. It already had a debt of about 80 billion dollars, to which about 700 billion dollars, as war damages, would have been added after the war. With the end of the war, Iraq has not been able to start a rebuilding programme because of the UN embargo which is persisting as Saddam has not been cooperating with the UN forces.

Fourth Largest Army and Sixth Largest Air Force Destroyed

In this war, about 4000 main battle tanks, out of the total 4200 deployed in the Iraqi battlefield, were either destroyed, confiscated or crippled; 9000 artillery guns and 2000 armoured combat vehicles also met with the same grim fate. Iraqi Navy was completely washed off. 259 Iraqi Air Force aircrafts were either permanently damaged or flown to Iran and have never been returned : 33 fighter aircrafts and 7 helicopters were destroyed in the air combat, 81 aircrafts were destroyed in the underground bunkers, and 137 found a rusty refuge in Iran. All the weapons and explosives were either spent or destroyed. 400 underground reinforced bunkers, out of the total of 600, more than 40 bridges on the Farat (Euphratese) river suffered permanent damage. Most of their 700 surface-to-air missile systems, NBC weapon factories, and 'command, control and communication' installations were completely razed to the ground. About 50 to 100 thousand soldiers[1] were killed and at least 50 to 60 thousand were captured as war prisoners. Quite legally, after their victory, the MNF have insisted for payment of war damages worth many hundreds of billion dollars.

Kuwait has regained freedom. The West is assured of oil flow. The USA need not worry about its place in the Middle East: No nation can challenge them. Saudi Arabia can feel safe, so also can Iran. Syria and Israel can also feel happy. Only Kurds are still being massacred by Iraq. And the dream of new 'World Order' which was fed to the world is still a dream.

UNO : A Theater for Big Players ?

Did the Iraqi public really want to incorporate Kuwait in Iraq, regardless of the will of Kuwaitis ? Has a smaller neighbour, which has a long historical past and is also a member of UNO, a right to exist as a sovereign state ? Freedom has been restored to a small state, but can we expect such a humanitarian action in future also from UNO ? But there can be no doubt that Iraq's attack on Kuwait was totally unjustified and the UN Security Council, perhaps for the first time, showed sufficient and commendable interest and passed speedy, appropriate and prompt resolutions and authorised an appropriate action plan. Although, it remains an open question if the UNO has really woken up to its responsibilities or does it remain a theatre for big players only. In the nineteenth century or the earliar half of the twentieth century such a justiciable action would not have been thought of. Such an action in the late twentieth century does indicate or seems to indicate that we have progressed towards a more civilised society. In the twenty first century, shouldn't we progress further !

---

1. Reliable data on this subject is not available.

To attack a small innocent nation and annex it, in modern times, is undoubtedly a sign of mean ambition. To enforce an effective control over such ambitions in a dictatorship is very difficult, if not impossible. Such mistakes are possible even in democracy but control therein also remains a distinct possibility.

Is The USA Also Somewhat Responsible ?

Now let us see the actions and decisions of Saddam from the strategic points of view. Having committed a blunder, a democratic country would perhaps have vacated the aggression respecting the UN directives of 2 August 90, giving due regards to the world opinion. An ambitious dictator would, instead, try to maintain his authority rather than save destruction. It is just possible that if the USA had issued a stern and clear cut warning, well before the attack, instead of giving temptingly (?) ambiguous signals, Iraq might not have indulged in such a misadventure. From this point of view, the omnipotent USA is also somewhat morally responsible for not having avoided such a large scale killing of human lives, destruction of property and indirectly causing misery to a whole nation. After all, the US does act as a police man !

History Teaches, Are We Learning

Saddam would have assessed his military strength vis-a-vis that of the USA. When the USA, from 7 August 91 onwards, started increasing its military presence in that area, Saddam should have given a serious thought at that time. He did give a serious thought and a decision he took, but strategically a wrong one. Saddam wanted to repeat Vietnam like situation, whereas General Douhet as early as in 1921, had said that no war could ever be won on the pattern of the last war, but with higher technology it could be. Vietnam could not have been repeated as the high-tech of the US had increased the range, lethality and precision of their attack weapons for aerial warfare vis-a-vis defence. Americans had learnt a lot from the mistakes they had committed in Vietnam War and also from the fiasco of the rescue operation for the US prisoners in Iran. Unnecessary restrictions like bombing in gradual steps to force the enemy to come to negotiating table, giving time to enemy for negotiations, not hurting the civilians etc. were, except the last one, not imposed on the forces this time. The precision guided weapons reduced the unnecessary killing of innocent civilians to a negligible number. The desert environment was no doubt adversely affected by the large scale bombing by MNF, but on the contrary it was Saddam who used environment-pollution as a weapon, thereby losing sympathy of the environmentalists' lobby.

Strength of A Chain is That of its Weakest Link

Saddam's completely defensive tactics could not have been successful against the high technology, although he could have succeeded in killing more MNF soldiers and in extending the war for a much longer period, if he had taken ground offence in the earliest stages. Though Iraq's air defence system was very strong, its vulnerable point was its weak ECM and ECCM. It is arguable that if they had sufficient ECCM and ECM, could they have saved their air defence radars from the stealth aircraft F-117 A, Tomahawk cruise missiles and SLAM missiles etc. (these were very difficult to counter). Even then, with effective ECM and ECCM, the war would have proved much more expensive and much more time consuming. The destruction that can be wrought by 56 F-117 As and 300 cruise missiles etc. would be severe but not like crushing the entire army. Air power would still have been necessary. If the ECM and ECCM of Iraq's air defence could work, MIG-29s and F1EQ-6s would have exacted a price which might have energised the US's anti-war lobby, thereby certainly creating uncertainties. Further, to prolong the lives of air defence radars, they should have, inter alia, put 'hot dummy radars' in large numbers. In addition, during such attacks, it was essential to deploy their interceptor aircrafts for air combat after considerable practice under thick ECM environment. It was vitally important for Iraq to have built its ECM and ECCM strength commensurate with the rest of its defence force. The MNF Air Force jammed Iraq's ground as well as airborne radars, but Iraq could neither develop ECCM to counter the jamming nor jam their radars. Iraq was not having sufficient strength in electronic warfare, although they had the largest and the most powerful SAM and AAA based air defence systems amongst the non-weapon producing countries. Although, one can discuss many ifs and buts for any war, this point is relevant because if Iraq had strenthened its EW capability, the story of this war would have been different. Obviously, Iraq had not realised so. The resources to acquire in-house developmental EW capability would have been only a small fraction of what Iraq must have spent on building of nuclear capability. Iraq could have procured 10% less of SAM systems and with these resources it could have built its EW capability. But it is easier to see solid weapons which produce fireworks rather than invisible weapons, however effective the latter be. As the proverb goes, the strength of a chain is decided by the strength of its weakest link and hence the MNF could break the backbone of Iraq's air defence system by destroying its radars. MNF attack missions, whenever required, either carried electronic jammers or radar destroyers (Wild Weasel) to easily and surely achieve success in their missions.

When Iraq lost its air superiority, how could its army have saved itself from the thunderbolts pouring from the sky. This was the vision of the earliest air-power theorists which had come true. But these questions did not bother Saddam as he was over-confident; and, we suppose, he would not have accepted any advice, sage or otherwise.

No Substitute to Competent Generals

Probably there was one tactical move through which Iraq could have achieved a bargaining stance. During the month of August 90 itself, they should have quickly captured the Saudi oil wells and some of its air bases by launching pre-emptive attacks to gain bargaining power. And subsequently they could have withdrawn their forces from Saudi Arabia and Kuwait by putting certain conditions e.g. Iraq be given Rumalhah oil fields and Bubian island, the Palestinian issue be resolved quickly etc. This move would have given Saddam a chance to honourably withdraw from Kuwait and still remain a force to reckon with. Though debatable and doubtful, such a tactics can be considered as one of the viable possibilities. This demonstrates the point we are trying to make. Don't let enemy build his strength when the war is about to begin or has already begun.

Capture of certain airfields of Saudi Arabia, which did not have sufficient number of airfields, would have deprived the MNF of these bases and hence reduced the effective air power that they could support. Alternatively after the resolution 660, he could have withdrawn from Kuwait showing respect to the world opinion, but held on to the Rumalhah oil fields and Bubian island. He would have gained considerable sympathy. But Saddam did exactly the opposite and so much so that he compromised the mobility of his army by deploying it in trenches, behind jungles of mines and barbed wire fences. Within a week of the start of the war, MNF Air Force achieved complete control over Iraq's air space. Indeed, Iraq was left with no alternative except that of receiving a big thrashing. In addition, the Scud missiles were not able to incite Israel sufficiently. Saddam should have taken steps to withdraw from Kuwait on 15 January 91 or even on 20 January 91. Withdrawal he made, but after receiving full 43 days of thorough beating and suffering mass destruction. By being pragmatic, Iraq would have saved itself from such a paralysing calamity. It seems, Saddam did not have sufficient able, mature, bold and outspoken Generals to advise him because such Generals were either shifted to political posts or were retrenched from time to time. In the early stages of Germany's attack on the USSR, during WW II, despite tremendous numerical superiority of the USSR in weapons, specially aircrafts and tanks, the USSR was losing battles after battles, because there were no competent Generals left to plan the war's strategy and tactits to deploy those expensive 'toys of war'. Did Saddam learn a lesson from Stalin's mistakes?

Air Superiority Reduces Losses Also

On the other side, what did the MNF lose ? They lost merely

43 aircrafts due to missiles, AAAs and small weapons and 21 aircrafts due to other reasons like mistakes having occurred during training, transport operations or engine failures. Only 162 soldiers were killed, 56 were untraceable and 13 became prisoners of war. Considering such a large magnitude of the MNF operations, these losses are extremely negligible and have established a record in the history of wars. Probably the rate of loss of soldiers during the war was less than the rate of loss due to automobile road accidents !

Targets That Got Away

Indeed, there were some MNF targets against which Scuds or 'Husseins, might have proved effective. Due to shortage of airfields in Saudi Arabia, a very large number of fighter aircrafts used to be parked in open. Saudi Air Force had 400-500 aircrafts to which 1000-1200 aircrafts were added resulting into severe parking problems. If all the 81 Scuds that Iraq had launched, were aimed at these airfields, Iraq might have destroyed 30-40 aircrafts, damaged 100-150 aircrafts, in addition to creating a problem for the MNF by disturbing their smooth and planned air operations. Further, the missiles that would have missed the aircrafts, might have damaged the airfield's vital services like air traffic control, communication centres, fuel depots, armament depots etc. Other possible targets would have been aircraft carriers which are very large in size and are densely packed with aircrafts, ammunitions and fuel. It appears that either Saddam did not have able military advisors or they gave Saddam only that advise which pleased him. Possibly, Saddam was sure of provoking Israel by his Scuds and thought that if Israel retaliates, the unity of MNF would be broken. He underestimated the wisdom of Israelis, but all these blunders at what human cost ?

Mistakes By MNF

It is not that the MNF did not commit any mistakes or that there were no shortcomings in their operations. Let us consider how they planned and coordinated 3000 air sorties every day. In such a short time the complete planning of the large scale assignments would have been possible only with the help of a high-tech information management system, the software for which had just been developed for European battle scenario. Since the contribution of Navy in the European battle scenario is negligible, the format of that software was not compatible with that of the US Navy. Due to these limitations, the instructions from Riyadh based command control centre, called 'Black Hole', were sent to the Navy through air-couriers whereas to others by computer-computer communication. When the instructions were passed through a messenger, apart from waste of time and efforts, it deprived them (Black Hole and Navy) from having an instantaneous two-way communication/interaction capability. No wonder, the US Navy felt hurt in that efficient and optimum use of their aircrafts, which were in significant number, could not be made due to lack of this co-ordination.

Although heavy attacks are required but it should be ensured that the weapon resources are not wasted. For example some of the air-to-surface IR guided Maverick missiles, costing $100,000 each, were used to destroy even fuel tanker vehicles. Many a time high-tech weapons were fired in the war without a justified requirement. The situation, at times, appeared to be akin to various models displaying their costumes in a fashion parade. The only difference was that the US was not "displaying" only to sell his weapons, but also to impress the world with his treasure of armaments.

LANTIRN (Low Altitude Navigation & Targeting Infra-Red by Night) system could not be made compatible with F-16. The MNF did not have enough knowledge about mobile Scud units. The army, for their attack missions, most of the times, though it is not unusual during a war, found the available information inadequate or even faulty which resulted into considerable waste of ammunition and time.

If management of the war is considered in totality, these shortcomings or inadequacies may be considered to be negligible and pardonable. But there is another side to look at. The pressure or tension, which the MNF had, during war after the air supremacy was established, was not the creation of the enemy. The tension was due to the aim of concluding such a complicated and huge war in the shortest possible time with smallest attrition rate. With this in view, the above mistakes did not have significant adverse effects. In addition, if the enemy could exert pressure, somehow like the mobile scuds did,many more mistakes would have surfaced. In fact the overall efficiency of Operation Desert Storm had bccn more than what is visible, despite the astronomical expenses incurred by the MNF. However, in no way such astronomical expenses reduce their glory in attaining 'the grand success' in the war.

Information War

A word about 'Information War' must be said. During a war, the warring country does not want to part with information which is not helpful to it and wants to give wrong information to its advantage. The 'Press' on the other hand wants to get the facts. In this formation war, the 'Press tought feebly & lost miserably, and the MNF comeout with flying colours.

Strength of the MNF

The reasons of their success in establishing air supremacy and winning the ground battle in one hundred hours could be summed-up as:

    i)   High technology : weapon systems, information management and electronic warfare,

    ii)   Appropriate strategic planning and execution,

    iii)   Very thorough training and preparations, proper organisation for gathering of information, sifting and use thereof,

    iv)   Maximum use of air power and hi-tech weapons,

    v)   Command & control - orderly, single point, autonomous and with minimum political restrictions,

    vi)   Higher caliber of soldiers, maintenance organisation and abundance of resources,

    vii)   Heaviest possible relentless attacks,

    viii)   Noble, clear and steady aim, mobility, surprise and high morale.

General Douhet's Prophecy Comes True

It took 70 years for General Douhet's statement (already quoted earlier) to be proved in reality, though, up to certain extent, its proof was visible in the Six Day (Arab-Israel) War and then in the Becaa Valley War (Israel & Syria). The Gulf War, for the first time conclusively proved that provided certain conditions can be met, a war can be won by air power, of course, to pluck the fruits of victory, the army would have to enter the enemy territory (without firing at the enemy !!) to take the defeated soldiers as prisoners of war and capture the enemy's land. Marching of army into Kuwait was totally different from marching of Allies' army into Germany and Europe in the last phase of WW II or for that matter in any previous war. In fact, in this war, this role of army of gathering the demoralised enemy soldiers can not really be termed as fighting a war, it was only winding up the show. In the Kuwait battlefield when Major Baker of Queen's Company of Grenadiers (part of Britain's Fourth Armoured Brigade) was asked to take the surrendering Iraqi soldiers as prisoners, he said, "I feel as if I am running a glorified taxi service". After a short while Brigadier Kardingli, Commander Fourth Brigade, cautioned all his soldiers not to open fire on the enemy but fire only warning shots.

Victory Through Air Power Through Electronic Superiority

Accepting the absolute necessity of army in any war, it would still be fair to conclude that the victory in the Gulf war was attained by the air power through the electronic warfare, the high-tech weapons and information management. Army is always required to march into and capture the land if that is one of the aims of the war. By saying so neither we are undermining the importance of the role played by the army nor the planning of General Schwarzkopf; we are only highlighting the decisive role played by the high-tech air power in this war and its development potential for future wars. From this analysis we do not wish to make a prophecy that all future wars would also be won by air power. We hesitate to establish such a principle for future, from the events in the Gulf war, because Iraq's capability in the vital field of electronic warfare was far inferior to that of the MNF. If these capabilities were balanced on both sides, the war would have been significantly different, for one, it certainly would have taken longer time, and second, the attrition rate of the MNF would have been higher, though, finally, Iraq would have retreated and vacated its aggression. If the battles are not quick and decisive, and the war gets prolonged, many other issues complicate the outcome. Having said all this, the conclusion that we have drawn i.e. the victory in the Kuwait war is the victory of air power through high-tech weapons, information management and electronic superiority, is still valid.

Air Superiority Essential for Victory

The only principle that can be established is that, now onwards for victory in a war the air superiority (including space) would be an essential condition. It has generally been accepted so far that 'victory is not possible without adequate air power. Our conclusion would go a step ahead in stating that the air power not only helps in making it difficult for the enemy to win but also is an essential element in achieving victory over the enemy. In turn, the air power gets its strength from the high-tech weapon systems and electronic warfare techniques, as has been demonstrated during this war as well as some earlier wars already discussed. The hi-tech weapons like Tomahawk surface to surface missiles, do not form part of conventional air power and are therefore put into a separate category 'Hi-tech Weapons', as these can be launched from ground or ships.

Now a question arises, if the victory could be achieved through superior air power, why did it take 70 years in establishing the statement made by General Douhet, when all the world over, air power was being given high importance. Apparition of radar and radar controlled air defence weapons, that Douhet could not have imagined, have delayed the materialisation of Douhet's prophecy. For the air power to be effective, relative safety of bombers and precision aiming capabilities are very essential.

Let us discuss the subject conceptually. For example, to destroy 10 bridges or radars, 300-400 ordinary bombs may have to be dropped, whereas only about 11 or 12 precision guided bombs or anti-radar missiles would be adequate for this task. With reasonable air superiority for dropping of 300 to 400 ordinary bombs about 100-200 sorties would be required whereas only 6 or 7 sorties would be sufficient for dropping 12 precision guided bombs. However, if reasonable safety is not provided, the number of sorties necessary would be 2 to 4 times the number already given. Based on this, by a rough estimate it could be very safely stated that the amount of damage done in this war by 1,15,000 sorties would have required more than 2-2.5 lakh sorties without the help of electronic warfare and high-technology weapons; time taken and even the casualties of MNF aircrafts and pilots would also have increased ten to twenty times or so. It is the supremacy in electronic warfare that permits the dropping of high precision bombs with accuracy and impunity. The present war expenditure, estimated as 50-80 billion US Dollars, in the other case would have been at least 100-150 billion US Dollars.

Shelling with ordinary bombs is waste of the available resources not only in terms of time and money but also in terms of guaranteed damage caused. Also, if destruction of a target 'A' depends upon the destruction of target 'B', after sending a sorty for destroying 'B', another sortie may be required for its confirmation if a reliable photograph could not have been taken in the first attack sortie itself. Therefore further tactical moves, which are based upon the destruction of the marked radars/ bridges, can not be executed with speed and confidence.

The capability of such precision aiming has been achieved on account of tremendous self confidence generated by ECM umbrella and due to electronically controlled precision guided bombs and anti-radar missiles which can destroy dangerous targets like radars reliably, efficiently and safely. This strength of airborne weapons made the mobility, firepower, and the range of aircrafts more effective and the electronic warfare really functioned as a force multiplier, par excellence. The term force multiplier indicates increase in the force in quantitative terms. The EW works in the qualitative domain, meaning it achieves something that can not be achieved by mere numbers. It permits an air force to play its full role and prevents enemy's air force to play their full role.

The actual fight or combat or raid is carried out by various weapon systems, however much they be dependent on electronic's magic. Electronic Warfare tactics or strategies can not win a war just by themselves but without their help air power also can not win a modern war. The role played by the stealth aircraft, the epitome of ECM and of high-tech in this war, clearly demonstrated the supreme capability of air power, when supported by electronic warfare.

Offence Vs Defence

At the present state of technology , it is the offensive which is stronger than the defensive capability, this is specially valid for electronically controlled weapons and electronic warfare. But, as has been happening during the history, tomorrow if defensive capability becomes stronger than offensive capability, the wars would be prolonged again. But day- after the situation may yet again get reversed. This 'cat and mouse' or 'shield and sword' syndrome is in the nature of war, only the introduction of electronic warfare, which is, apart from being a high-tech equipment, a game of wits, has made this syndrome rather sensitive and liable to change at a very much faster pace than in the past. This phenomenon was conclusively demonstrated in the battle of Britain of WW II. One of the reasons of high attrition rate of US soldiers in Vietnam war was the strong EW capability of Vietnam, though provided by the USSR. Innovative mind with the help of appropriate technology can tilt the balance of EW, either in favour of offence or defence.

Harnessing of Air Power :

The air power has to be adequately harnessed for which General Schwarzkopf and General Horner had taken appropriate steps, and they deserve special appreciation. The steps are :

1. Training and battle exercises during the Desert Shield operation under a realistic environment.
2. Selection and deployment of appropriate weapons and electronic counter measures.
3. Perfect selection of targets and their sequence.
4. Heaviest possible attack missions to ensure destruction of targets so that the enemy does not get any time to recoup and attacks can be planned with much more confidence.
5. Quick assessment of war damages.
6. Appropriate deployment of all available resources, i.e. proper concentration of forces.
7. Excellent planning for about 3000 sorties per day, their execution and proper decentralisation.
8. Very effective use of electronic warfare and management of sea of information and disinformation.

Timing for Ground Attack

The MNF had already planned to conduct the ground attacks by army only when the Iraqi Army strength in Kuwait battlefield was to be reduced drastically and all the land routes to Kuwait for obtaining help from Iraq, were to be damaged. This had to be so because the MNF Armies' strength was three fourth of that of the Iraq's Army, whereas it should have been 3 to 5 times if ground offensive was to be resorted to in the early stages of the war. Seeing the reports appearing in newspapers and magazines during that period, the MNF never gave such statement openly. They went on giving statements that they were fully prepared for the ground battle and may flag it off any moment. They were trying to shield their army's weaknesses by such a propaganda, not an unusual activity in a war and quite justifiable.

Modern Technology and Air Power

Therefore, another lesson, if are has to survive in a war today the nation must be rich !! If during the first three days of the war, 200 surface-to-surface guided missiles Tomahawk and a large number of stealth aircraft F-117A would not have undertaken intensive, extensive and precise attack missions, and if the air force have had to launch attacks employing other weapons including electronic warfare, the war's progress would have been slower but the outcome would have remained the same because air supremacy would have been achieved though somewhat belated.

It was most essential that a few important NBC factories, Scud missile systems, airfields, air defence system, 'command, control & communication' and radar systems were made ineffective or destroyed. It is not just a chance that the first three targets of the MNF on the night of 16 January 91 were radars and communication centres. During the war, Wild Weasels, the ECM killer aircrafts, fired about 1000 HARMs whereas the Tornado GR-1s launched about 100 ALARMS-advanced anti-radar missiles. With the help of precision guided weapons, ARMs (Anti Radiation Missiles), and ECM escorts, the MNF first established air superiority and then air supremacy. It was this ECM generated air supremacy that permitted even the grand old bomber B-52G, brought back from retirement, to quickly and economically liquidate, both physically and psychologically, the Iraqi Army by pouring old conventional bombs from the skies. Out of the total bombs dropped in this war, 30% were from the grand old B- 52Gs and only 9% of the total ammunition was with precision guidance which were responsible for handling the most important and difficult targets. On the other hand, it is also worth noting that the percentage for destruction of targets was in the reverse order viz. about 30% targets were destroyed by PGMs and about 9% by B-52Gs.

Indigenous EW Capability Essential

Another extremely important and indeed vital lesson comes out from the impotency of the expensive, modern and hi-tech weapon systems of Iraq. After all Iraq was an oil rich country and Saddam was generous with purchase of modern hi-tech weapon systems and so were the Western powers to supply him what ever he wanted. The hi-tech weapon system of Iraq had ECM and ECCM features, that could be sold only to a very rich country. Then how come the impotency, the abject inaction from the Iraqi Air Force and their air-defence system ? As already explained, ECM- ECCM is a dynamic and ever evolving game of war. Even the mighty MNF under the leadership of the super power USA, kept on upgrading their EW capability, so much so that they pressurised France to give them Mirage F1, its armaments, Rolland SAM system with all intricate details to enable the MNF to develop more appropriate ECM to counter the existing ECCM of the French weapon systems. They installed the latest jamming system in F-15s almost on the 'D-day'. Whereas Iraq did not develop any ECCM, why ? Simply because Iraq did not have any indigenous capability to carry out any improvements in the existing ECCM capability. One of the most important lessons is that along with hi-tech weapon systems, a nation must have indigenous EW capability good enough for real time generation of superior ECM and ECCM, to remain fighting fit in a modern and dynamic hi-tech war.

MNF possessed predominantly more high-tech weapon systems than Iraq. Specially a superior inherent EW strength capable of real time development. They also had developed more expertise for night attacks for the obvious reasons of better safety and that is why most of the difficult targets were attacked during nights. Even with full jamming of radars - the electronic eyes, and radar controlled weapons like SAMs, day attacks were considered dangerous because some retaliation is still possible with the aid of human eye. If Iraq had a strong EW capability, the air war, per force, would have been reduced to an 'eye to eye' air war. Then superiority of platforms and pilots would have determined the course of battles.

Expensive Weapons Favour Rich; UN Favours ?

Another important point is that all these high-tech systems like F-117A, Tomahawk, Patriot, HARM, ALARM, Maverick and space satellite systems are exorbitantly costly and none other than the most rich and powerful nations can perhaps afford them. The future high-tech weapons would be even more expensive. Would the future wars then be one sided affair ? It sounds logical to answer the question with 'yes'. Therefore, does a need for 'civilised countermeasures', against such predominance, not become more appropriate and sensible now, even more than in the past ? Should not the UN become 'democratic' in the real sense? Should it not have better representation, more representative restructured organisation and methodologies to handle various crises ?

Do Radars Have a Role in Future ?

The outright success achieved on the strength of electronic warfare in establishing the air supremacy puts a question mark on the future of radars. Some analyst feel that the grand success achieved by the ECM systems, especially by F-117A, HARM and ALARM, threatens to drop the radars in a garbage dump in future. However, such a situation is normal in competitive development of war related technologies. One develops a shield that threatens the effectiveness of swords and then the other side develops a sword that makes shields ineffective, and so on and so forth. As a counter to the above ECM systems, development work on suitable ECCM systems would have already begun in a full swing and undoubtedly they would succeed. Developments of low level interception radars and radars to identify and track stealth aircrafts have also been taken up on top-priority. Already Russia is claiming to have developed a super air defence system which can detect and track even stealth aircrafts. In fact, now one should give due importance to high-tech 'electronic supremacy' that makes an essential and major contribution in establishing 'air superiority' and even 'land superiority'. It should be so because, most of the modern weapon systems are dependent on electronic's magic.

Since the ECM systems succeeded in the Kuwait war, one should not conclude that the radars have become (or would become) redundant. The MNF aircrafts undertook about 1,15,000 sorties, out of which about 60,000 were on attack missions rest being supply, transport or logistics missions and they dropped a total of about 90,000 or one lakh tons of ammunition. It means that the MNF could achieve the average ammunition dropping efficiency of only 1.5 tons/sortie whereas their aircrafts had the ammunition carrying capacity ranging from 2 tons/sortie (F- 117A) to 15 tons/sortie (B-52G). For reducing the ammunition dropping capability of the MNF aircrafts so drastically, the credit solely goes to the Iraqi air defence network of which radars were an integral, essential and vital part. Therefore radars carried out their performance to the best possible extent, it was the ECCM that was lacking which caused their failure. Even then, it is not fair to state that radars without guaranteed ECCM have become out of date because, even under such unbalanced conditions (as in Kuwait), the radars were responsible for reducing the average bomb load of the MNF aircrafts say approximately from 6-8 tons per sortie to about 1.5 tones per sortie. With appropriate ECCM, radars again would have as important a role to play in future, during peace and war, as the human eye has in life.

Is Tomahawk too Expensive for its Precision ?

The famous defence analyst, Professor Edward Lutwak, opined that Tomahawk is a 'hopeless' missile. His main argument is that, with 1100 km range its aiming accuracy of 6-10 m is rather poor, specially in the context of its price and the effectiveness of its warhead. However, we would not conclude that it cannot be effective in a war though everybody wants that his weapons should have pin-point accuracy and cause substantial and irrepairable damage to the target and should not be expensive. The present aiming accuracy of Tomahawk will make it less effective for targets smaller than 10 m in size or for those targets which are required to be hit with an accuracy better than 6-10 m. But in a war, there are many important targets which are larger than 10 m and do not need to be hit with 6m accuracy. And if, for any specific and important target more powerful warhead is required, more than one Tomahawk (with increased cost) could be launched to achieve the desired destruction. The important aspect to assess is how important the target is and which weapon would be the most cost effective to attack it and how cost effective are the alternate weapons. The cost estimation also includes the possibility of losing the weapon-carrying aircraft and even its pilot if a bomber aircraft is the only other alternative for attack. What weapons would be most appropriate for very critical targets, heavily defended by lethal SAM systems, with minimum loss of own soldiers, in the first attack when even jamming may not be reliable enough? Answers would be Tomahawk, SLAM or F-117A in to-day's technology, if one can afford them.

In this war, USA specially wanted to attack at a very fast pace so as to keep the possibility of losing pilots the lowest and hence during the initial war conditions, when risks were highest, Tomahawk proved to be one of the most cost effective weapons for highly critical targets. During the later stages of the war, when the air supremacy was achieved, Tomahawk was sparingly used, even though they were available in plenty. There is no doubt that one of the factors that brought about the air supremacy so fast was this Tomahawk itself. What makes Tomahawk difficult to use is the necessity of digital maps of the target areas, which may not be known well in time. Therefore good forward planning is essential for being able to use Tomahawks.

Work has already started to improve the accuracy, range and fire power of the Tomahawk missile like for any other attack weapon. There should be no doubt that it proved its usefulness in the Kuwait War and therefore has reserved its berth (in an improved version) for the future wars also. It is worth noting that such surface-to-surface missiles might give a new direction to the future wars specially if their cost can be brought down which may not be too difficult. Now the next development to watch in this direction is a suitable anti-missile to counter a Tomahawk. Would it be waste of 1.2 million US Dollars, if a successful anti-missile missile is developed? No, not really. Then tactics, surprise etc play their role, also, how many critical targets would the enemy be able to guard with its very expensive anti-missile missile systems! And, in any case, cat and mouse game goes on.

Do Stealth Aircrafts Have A Future

The success of the stealth aircraft F117-A, in the Kuwait war was, no doubt, 'par excellence'. Indeed the bombing role of an aircraft, that dominated during WW-II, became obsolete on development of superior fighter/interceptor aircrafts. Now that role of a bomber aircraft has been brilliantly resurrected by F-117A.

Despite the phenomenal success of F-117A, a modern military thinker has thrown a question, " Do stealth aircrafts have any future?" One of the foremost experts on tactical weapons, Professor Kostas Tsipis feels that the frightfully expensive stealth aircrafts would not prove to be successful as their low speed can land them into a dangerous trap of the enemy. It is true that compared to a modern fighter aircraft, the speed and manoeuvreability of F-117A is quite low as, after all, its shape is made of plane surfaces and not aerofoil surfaces that provide the optimum lift. However if the stealth is effective i.e. if radars and IR systems can not detect it and no human eye can see it during night time, what dangers has it to face? This is the only military aircraft in the world which has no provision for a weapon on-board to save itself, neither artillery guns nor air- to-air missiles. Being 'invisible' itself is its safety. If, in future, a radar or IR system is developed which can detect it, some more of the surfaces of F-117A would have to be made of composite materials rather than duraluminum. As the technological developments make headway to detect the stealth, so would the techniques to increase their stealth capabilities to counter it. Russia is already claiming to have developed a radar which can see aircrafts like F-117A. Such evolutionary research & development activities would go on and the relative capabilities would always be continuously changing. New tactics and ways of generating surprise etc. would also be developed. Of course stealth aircraft would become even more expensive, thereby remain beyond the reach of all, excepting the most rich nations.

No doubt, F-117A is not a weapon system for poor, like AK-47 is and perhaps. AK-47s can do more damage than F-117A, provided the victim nation does not understand the impact of terrorism inspired and supported by an enemy and does not know how to retaliate.

AWACS Will Keep Awake

AWACS[1] (Airborne Warning And Control System) aircraft proved to be very useful for the air defence control function, to safely detect, control and provide directions to about 3000 sorties per day in this war. It is noteworthy that three AWACS aircrafts were always airborne to keep a round-the-clock vigil on the complete war theatre, in addition, another AWACS was always in the air to come to their rescue immediately, lest one of them should develop some fault or get busy in refuelling. Though it must be noted that in this war AWACS got very limited chances to keep a vigil on the enemy's aircrafts, nor was there any necessity to defend them (AWACS). In a war AWACS type of aircrafts, because of their importance and vulnerability become prime targets. However, in this war the AWACS did provide the information regarding the location and direction of enemy and their own aircrafts and took over the control of the situation whenever it was necessary. It remains to be seen how effective would AWACS type of aircrafts be under adverse conditions. But it is reasonable to believe that given enough air superiority, such aircrafts would be vitally useful.

Due to the overwhelming success of the ECM systems of MNF, the Iraqi aircrafts could not really fly and hence MNF had air supremacy from the beginning itself. Thus to draw a conclusion from the evidence in Kuwait war that AWACS is very useful is of limited value because it operated only in the air supremacy environment or in the absence of enemy's ECM. However, in a situation of doubtful air supremacy or dense ECM environment, the airborne warning and control aircrafts would

---

1. AWACS - has been used here in a generic sense, though it is the name of a particular type of aircraft, EA-3 sentry, possessed by the US, which is used for this purpose.

be in severe danger, as they would be prime and comparatively easy targets. They would have to run for their life when attacked, despite being defended by fighter escorts, and if they are jammed then they would be of little help. The advantages of AWACS in such an environment could not be ascertained in this war. It is noteworthy that the Iraqi AWACS type of aircraft could not make any sorties excepting one that too when they fled to Iran. After all why ? Because air superiority is a vitally essential condition for such an aircraft to fly reasonably safely, for which in turn it is essential to have a modern strong air defence system along with modern interceptors. A point to note is that AWACS helps, inter alia, in strengthening air superiority if there is enough air superiority for it to function. Also, AWACS is a weapon system of a rich nation.

RPVs Fly Into Future

The remotely piloted vehicles (RPV), Pioneer and Pointer, proved their capabilities beyond doubts, not only in this war but also for future wars. If a piloted bomber is the most effective weapon, considering all the conditions, at times, RPV may be better because risk of a human life is not involved. RPV has potential to be one of the most important weapons of future. Again, RPV is waiting for an appropriate technology to find its dominant place in the galaxy of weapons as it is either too cost-ineffective or not reliable enough at present compared to its alternatives that are available.

Quality of Soldiers Decisive

After suffering a setback in Vietnam and Operation Desert-1 (operation to free the US hostages in Iran), in addition to other lessons, the US discovered that the basic competence of their soldiers was rather low and it needed rectification. With improving technology and weapon systems, the quality of men to handle them must also improve. As per Les Aspin, the Chairman, US House of Representative Committee on Armed Services, the pay of an American soldier at the time of entry into the service, nine years back (early eighties), was 84% of the national minimum wages; and only about half of the recruits had high school qualification. To attract better quality into the armed forces the salaries of the recruits were increased, they were encouraged to study and gain knowledge, efforts were made to increase the living standards of their families, and more competent recruits were given bonus to get enlisted. By June 91 about 95% recruits had high school qualification and 100% recruits passed the departmental examinations, conducted by the Services. The discipline related problems had reduced. The high calibre US soldiers (army, navy and air force) no doubt played a major role in winning the Kuwait War in a short period with record minimum rate for war casualties and super maintenance of hi-tech weapons.

Single Command and Control

Les Aspin had also stated that in Vietnam War there were two Armies and four Air Forces with their independent chains of command and control, which consequently resulted in many problems. In Kuwait War, the command and control was with one Supreme Commander, General Schwarzkopf. Full powers were given to this Supreme Commander by the US President, Defence Secretary and the Chairman, Joint Chief of Staff. This resulted into an unprecedented cohesiveness in the MNF and consequently all the forces could work in unison. To obtain unity of command is more complex and difficult than it appears. It is also a question of culture. Firstly, 'politics' within the military must be discouraged. Secondly, the higher authorities should not encourage back-biters and thirdly, the entire military system must aim toward fairness and justice and also must appear to be so. These qualities for development need intelligence, professionalism and character everywhere, specially so, at the higher echelons. Firmness of aim is one of the important principles of war, single command and control is equally so.

Good Maintenance is a Force Multiplier and Morale Booster

It is essential to mention the important role played by the maintenance and supply system of the MNF which was the backbone of all the Operations in Kuwait war. Its significant contribution is evident from the fact that about 2000 aircrafts made about 3000 sorties per day for about 2 months, as per the dynamic requirement of war and that too in a hot sandy desert environment, probably the worst environment for aircraft performance and maintenance. The supply of spare parts and sub- assemblies etc., within 48 hours of demand, from distances of 1500 km to 8000 km, first at one place of the battlefield and then in the whole of Saudi Arabia, was a very difficult and challenging task. In the absence of an efficient supply and maintenance cover, it would have been impossible to undertake 3000 flights per day, each of 3-4 hours durations, on an average. Apart from the war material, think of the amount of food, water and petrol requirements to be met on time for the Sixth Armoured Division (US/French), the Eighteenth Air-borne Corps (US) and the Seventh Corps (US/British) for their 100 to 300 km long outflanking move beyond the Saudi border, under extremely treacherous desert conditions.

The deployment of high-tech computer systems for planning 3000 sorties per day had been extremely important and hence their (computers') maintenance was also very critical. If the maintenance was not so effective, the same 2000 aircraft might have done only about 2000 instead of 3000 sorties per day, a reasonably good figure under severe conditions of the sandy and hot desert. Therefore maintenance certainly acts as a force multiplier. In addition, during the war period, to suit the new assignments the old systems were suitably modified in the battlefield itself. Many US pilots, in their interviews published in different magazines, had all praise for the up-keep and maintenance of their aircrafts during this war. This kept not only the aircrafts but also the fighter pilots' morale sky high. Therefore, is not the morale of the maintenance personnel even more important ? Morale of U.S. technical personnel was, no doubt high, but quite a few militaries in the world do not realise it. With increasing role of technology and more so of high technology in a war, the maintenance must be given proper importance.

Logistics by 'Desert Express'

All the material, especially the necessary spares for repair and maintenance, were made available within 48 hours. They were airlifted to the actual battlefield from USA by 'Desert Express' aircraft. This itself was a great achievement. Providing right spare in the right place at the right time is no doubt difficult, but only a part of the task. To complete the task, the specific spare part must be fitted at the right place in a complex machine, under difficult, demanding and may be dangerous conditions by maintenance personnel, after correct diagnosis. Major Jarnigan, a pilot of F-16 (157 Fighter Squadron of the US Air Force) told Mr. Lenorovitz of AWST magzine "We are extremely lucky to have such capable maintenance crew who could correctly tell how our aircrafts are functioning".

Information Management

Just as EW functions in the invisible domain, unlike Tomahawk or laser guided bombs, information management also operates in the abstract. Another difficulty, just as was experienced in EW, is very fast progress of information management from the 'horseback' to 'speed of light. It is also worth noting that advent of EW to interfere with radio signals started a race to develop new technologies for information collection and communication. Then tremendous increase in the amount of information available to commanders, due to new electronic sensors, forced integration of communication systems into information systems and then to C^3 - Command Control and Communication, and then again to C^3I - C^3 and Intelligence. 'Unless there is information management, there is information chaos', is an important lesson.

Preparation Time

The US and MNF got more than five months for preparations. They used it with dedication, wisdom and diligence and got themselves to a high state of preparedness. There is no doubt that if they had not got this 'preparation leave', and had not used it so well, the casualty rate of MNF would have been far more, time required to win the war would have been longer and many more problems would have been faced. Perhaps the US or the MNF were lucky to have got such a long gap for preperation or perhaps this long period also served a political purpose. Kuwait didn't get time even to call for help. Quite often, one is not lucky enough to get time for preperations. After all, 'surprise' is one of the principles of war. Training and preparation are, or ought to be, the main activities during peace time. A very important lesson that comes out is that thorough preparation for a given war is essential and that is why the saying, "the more you sweat in peace, the less you bleed in war". But there are many countries where during peace time 'Defence' is invariably in the tray marked 'can wait'.

Hi-Tech Creates Favourable Conditions

Some experts have commented that the Gulf War was fought under unusually favourable conditions. After the first few days of air combat, both the air and ground wars bore resemblance to structured training exercises rather than to intense combat. What such persons forget is that, it is the same, proper use of hi- tech weapons like information management, electronic warfare, F-117 A, Tomahawk, SLAMs etc. and exploitation of enemy's weakness that created the so called 'favourable conditions'. Hi-tech weapons with appropriate strategy and tactics in modern days, with their superiority, are capable of creating such 'favourable conditions' during a war.

Hi-tech information management (organising successful attacks with 2000 - 3000 sorties a day with aircrafts coming from hundreds and thousands of kilometers for attack; weapons like Tomahawk, which are capable of stealth attacks up to 1000 kilometers, F-117 A, stealth aircrafts, capable of penetrating any air defence etc.) provides a capability of fighting a war on any scale, specially so because of the inherent mobility flexibility and quick reaction capability produced by it.

It is true that one cannot draw those combat-lessons from this war that assume equality of the two sides, almost till the end of the war. There are lessons in this war that can guide us to create a military superiority, through air superiority, electronic superiority and information superiority when the two forces are not very different in their strength, on the whole, to begin with. It is important to note that modern hi-tech systems help in exploiting any weakness in the enemy's defence system and quickly create a 'superiority environment' and win the war, most cost effectively, with minimal human and material losses in shortest possible time. Such a 'blitzkrieg' was not possible untill WW-I, as a 'trench cum machine gun' combination could create almost insurmountable block for the progress of war, because role of technology was not properly understood by the warring nations. Also a hi-tech victory does not let you rest on your laurels, it knows it is out of date the day it has won a war. It should also be mentioned that technology is a bad master but an efficient, effective and obedient servant. How to master technology, specially the illusive, dreamy, foggy technology which may look like an innocent child and yet have the tremendous potential of a full grown man ? The US was one of the last countries to realise the importance and capability of an aircraft, although it invented the same. A team of experts and professionals viz. scientists, soldiers, engineers and military thinkers must study it, all the time, to look into future, both near and distant.

Environment as Weapon

Weapons, even excluding nuclear, biological, and chemical (NBC) weapons, by their nature cause long-term damage to the human environment. In any war, good or bad, environment is the definite and real casualty. This is bad enough, but use of NBC weapons, like dropping of atomic bombs on Hiroshima and Nagasaki, spraying of defoliating agents in Vietnam War to clear jungles and to use other toxic gases was undoubtedly a curse on humanity. It may be debatable that the atomic bombs dropped over Japan brought the bloody, cruel and inhuman war to an immediate end and thereby saved so many lives and so much destruction. Unlike others, these weapons, especially nuclear weapons, damage the 'present' as well as distant 'future'. Although there are international laws against the use of NBC weapons, a desperate and incompetent leader may resort to these when threatened with defeat. It is therefore desirable that NPT be made fair, universal, and accepted universally. NPT should not be used to maintain the superiority of some nations because with nuclear weapons, there would be no victor left alive to enjoy the booty of war however superior his nuclear capability. Although it may also be mentioned that the frequency and duration of wars have reduced after the advent of nuclear weapons, but apart from nuclear deterrence, this reduction is also due to impact of United Nations and also the fact that wars have become expensive and cost-ineffective.

Somebody should ask Saddam or must he question himself: What did he achieve from this war ? or for that matter, what did Alexander the Great, Napoleon or Hitler achieve ? I am sure Saddam would not retort, "Ask George Bush what has he achieved ? because George Bush would very gladly oblige, "Ask Kuwait, please".

Should we all ask, " Would 'Homo Sapiens' ever outgrow war ?" Indeed, is war not spreading its tentacles among communities of a nation, among different groups within a community, among families of a group, and among members of a family ? Actually wars must be fought, between good and evil, within the heart of every man, then only can all wars be won.

The instinct for survival, desire to enjoy freedom and desire to rule over others are surprisingly dominant. Though the USA has no nation to match its strength, the new nuclear policy of the USA , after the break up of USSR, still demonstrates this syndrome. No doubt, USA has reduced the level of nuclear weapons, has withdrawn all nuclear weapons from its ground forces, has stopped the day to day alert condition of the strategic bombers, has abolished the Strategic Air Commands etc., yet the basic policy of nuclear supremacy is being maintained and capability to develop further is also kept active. Whole world fears, what if some nation clandestinely develops and threatens to use nuclear weapons. With sensitive devices, any testing of nuclear weapons can be monitored. Can the world community not come to an understanding on non-proliferation of nuclear weapons ? If all the nuclear weapons on this earth are destroyed, the USA would still be head and shoulders above the rest in conventional weapons also. The USA, therefore has nothing to worry with a fair NPT. Is it this fear only as described above which is driving the USA or is it also to have a capability to strike terror in others that is responsible for such double standards. If it is not the fear of misuse by someone but the ambition to dominate that is driving the most powerful nation of the world, where are the ideals of liberty, fraternity and equality? But if it is the fear of misuse, an acceptable NPT should not be a problem that can not be solved.

To prevent Saddam from using nuclear weapon as a blackmailing threat, NBC weapon factories were one of the first targets to be attacked by MNF. And to prevent such an occurrence in future, as far as Iraq is concerned, efforts are still being made by the USA and allies to destroy Iraq's capability to produce NBC weapons.

As already mentioned, in any modern war, environment is a definite casualty, regardless of any other consequence, but it is Saddam who used environment as a weapon. Probably he thought, like many of us, that all is fair in love and war. But don't we know that 'ends should not justify means !' Anyway, Saddam threatened to burn the oil wells in case of defeat, and actually he did so. He also spilled huge amount of oil on the land and the Gulf, which caused immense and cruel damage to sea life, a loss of thousands of billions of dollars worth of oil, as well as the depletion of the energy resource which is in short supply on the earth. Spewing of carbon dioxide and other toxic gases caused immense sufferings to innocent human beings and irrecoverable damage to the environment by taking this earth a step closer to the 'green house effect' which is increasing the temperature of the earth. Such an increase in the temperature is gradually changing the climate including the rain pattern on the earth. It would melt snows from the poles and cause deluge, not mythological but actual. It was Madam de Pompadour who had said, "After us, the deluge " Saddam actually behaved so. Environment can be a weapon of only a dehumanised soul.

On the other hand, destruction of water and sewage treatment plants of Iraq by the MNF, though might have been considered desirable on military grounds (even here, many wise persons would disagree), can not be justified on humane grounds. The MNF had enough weapons to attack a large number of other types of targets to force Iraq to surrender within the same period. Such a destruction has polluted the two rivers, the life giving rivers to Mesopotamia, because untreated sewage is flowing in the rivers now for the past five years. The continued economic sanction is preventing Iraq from getting these plants repaired. Economic sanctions should not, and must not extend to repair of the damage of the environment, which belongs to the whole family of living creatures on this small spaceship, the earth. Such a polluted water is causing diseases like Jaundice, Cholera and even Polio, and the victim is the innocent public and the water creatures. International laws need to be created to prevent deliberate damage of environment during war, whether in attack or in defence, whether by NBC or other conventional means and then restoration of environmental damages ought to be carried out on priority. As it is, the powerful weapons cause immense damage to environment on the earth we live on, the water we drink and the air we breathe in, and all these cannot differentiate between a friend or a foe. When war expenses are estimated, it must include the money required to remedy the environment to the extent feasible and ought to be given top priority. In this war, the funds required to repair the damages have been estimated as about 700 billion dollars. Sadly, this does not include the remedial cost to pollution. Indeed, the environment should be the first item on the agenda for repair and then money should be given for other damages. Such a concern for environment is the concern for the family of men, for children of today and tomorrow, a concern only civilised homo sapiens are capable of showing. Are we men enough ? Are we 'homo sapiens' enough ?

The capability to go to war has certainly increased, in terms of size of army, air force and navy, of expensive and devastating weapons, of mobility, communication, navigation, surveillance and logistics etc. Indeed, the super powers can destroy this world ten times, if they merely wish to. But is it not a pity and biggest irony that such a tremendous increase in capability of production, technology, of making this planet a village, of travelling into space beyond the solar system, of training the man to do anything, is not only being frittered away but also being used for astronomical destruction. Imagine this hundreds of billion dollars of war expenditure (not counting the damage in Kuwait and Iraq) being used for peace, prosperity and health. Would not Iraq, Kuwait and this small planet have been a better place to live in ! As we have already said, technology has advanced to make this world a heaven, but have we ?

"The trifles of operational leadership were something that anyone could perform."

—— Hitler, in december 1941, stated to the then Commander-in-Chief of German Army, General Von Brauchitsch, while dismissing him and taking over the charge himself.

"War is such a serious matter that it cannot be left to Generals".
                                              - Georges Clemenceau
                                              (French Premier,1918)

"Thus far the chief purpose of our military establishment has been to win wars. From now on its chief purpose must be to avert them".

- Bernard Brodie

Chapter 11

Future Wars

Technology Has Advanced, Have We ?

Can we confidently say that the progressive man, who is about to usher in the 21st century can avert a war?[1] Cicero had said, "let wars give way to peace". And was it not Franklin D Roosevelt who had written in April 1945, "More than an end to war, we want an end to beginnings of all wars.?" Certainly it is a dream worth pursuing. Did the Kuwait War prove the necessity or futility of a war ? Indeed, in real and long-time terms, what did Alexander achieve, or Napoleon, or Hitler or Saddam or for that matter most of the wars ?

There would be many situations when negotiations would fail and war would be a tempting alternative to impose one's will or to seek justice. Many analysts have said that if the Western powers were not so keenly interested in the oil of Middle East, - USA probably would not have got actively involved to this extent. But they forget, if there was no oil, Saddam also would not have risked his entire nation to begin with. When has any nation taken a severe action, not mere

---

1.   It reminds of me of an old proverb : " Advise none to marry or to go to war". Well certainly one should not go to marriage as if one goes to war, or vice versa.

V.M.T

words, without having an interest! This is a controversial issue as USA itself is not as much interested in the oil from Middle East as in maintaining its influence over the region and its trade; and for that purpose in creating friends and helping them. At the same time, it would be unfair to take somewhat negative view of the USA's involvement in the war. No country in the world is as innocent as a white lily, then, why do we condemn the USA, even when it has taken up a cause which is noble also. The USA and the West had powerful vested interests in attacking Iraq and in decimating the military might of an ambitious dictator but at the same time justice was also done in restoring to Kuwait·its sovereignty which was usurped by too unscrupulous a dictator, simply for feeding his dream to become a regional 'Big Brother'. Even a devil must be given his due and therefore this tirade and abuse against the USA on liberating Kuwait is not easily understood. It might have appeared that the MNF was too severe in destroying Iraq, but the situation till date makes one feel that the MNF ought to have destroyed the remaining 6 (out of 24) nuclear facilities, that would, in addition, have made the process of normalcy faster one. Now again why only in this region, the USA might be aspiring to dominate the whole world. Just as Saddam's act was reprehensible, so would be this if the super power USA really tries to "dominate" and "dictate" the world through the use of its might.

Future of War in the Remaining Century

We ought not to be naive to believe that everybody wants to avoid a war, however cruel or inhuman it may appear to be. It would be realistic to admit that, 'to war is human & to avert it is divine'. As already discussed, there can be many causes for a war to take place. One of the major causes of war was abolished with the end of WW II, when "fight for colonies" or 'lebensraum' which was considered legitimate till then, was given up as unethical and impractical, and consequently many colonies were liberated within a short period. With the dissolution of the second super power, another important cause for wars has now vanished viz. fight for dominance between the two super powers, who have either provoked and fought many proxy wars in the second half of the century or have taken sides of warring nations.

Other causes like territorial disputes, historical residues of earlier wars, simple greed, ambition, religious or racial differences and enmities etc. still remain. Also remains the desire of the super power to maintain its dominant superiority , now that it has achieved it. Since the super power, fortunately a democracy, preaches liberty,equality and fraternity,it would be difficult, though not impossible, for it to use brutal force to impose its will on another nation, unless it can muster world opinion in its favour. In any case, it would not be required to use any nuclear weapon because of its immense superiority in the conventional hi-tech weapons also, except if the opponent, somehow, in possession of a nuclear weapon threatens to use it. Deterrence till a fair NPT is signed would remain, practically the best policy as far as nuclear weapons are concerned, though a mad despot could still use a nuclear weapon. (For nuclear weapons, the best policy, ideally speaking, is not deterrence but destruction of all of them.) It also appears that within next four to five years, radiation weapon would not become practical even for the USA. Therefore even if the satellites are used as war weapons, they would be used for communication, surveillance, navigation etc. Therefore, next war, if it takes place in near future, would not be with nuclear weapons nor with radiation weapons, but with conventional & high-tech weapons, on which improvements are being made all the time. But what is the likelihood of a hi-tech war in near future, not counting what is happening in 'Slav' countries now.

It may appear that the technology gap between the super powers and others, would only reduce with time. This statement and aspires may be taken as true for one always expects to catch up in the race of technology. Japan has done so in some fields, but an exception would only prove the rule. Looking at the growth of technology, we find that, generally , it is not linear but somewhat exponential i.e. in a given interval of time, the technology always moves faster than it had moved during the same interval in the past. Therefore, a leader would always be growing faster and faster than laggards. It is true that in the world of inventions, unexpected jumps are possible, but that would be an exception and only highly developed nations can really benefit from it. Unless the USA drastically slows its growth of weapon development and some countries manage to increase development and growth of weapons, the USA would continue to lead head and shoulders above the rest. Therefore no nation (Russia has still to solve its complex problems) can challenge the USA to war in the foreseeable future and expect to win, or to come out without being badly mauled. If the USA declares war on any country, the victim country should follow the adage, 'discretion is the better part of valour' and should make all attempts to gather the support of world opinion otherwise it should be willing to suffer the fate of Saddam Hussein or worse. As already mentioned, electronic superiority, air superiority and hi-tech are the three dominant conditions that had determined the outcome of the Kuwait war. There is no reason to believe that any other nation can defeat the USA· in these three domains in the near future.

Let us look at the cost of some hi-tech weapons (at 1990 level) that were used in Kuwait war. A Maverick missile costs one hundred thousand dollars and was used at an average rate of 100 per day. A SLAM missile costs one million dollar and many of them were used. A Tomahawk missile costs 1.2 million dollars and a few hundreds of them were used. An F- 117A stealth aircraft costs close to 120 million dollars, and 56 of them were deployed. A modern tank costs around 2 to 5 million dollars and several hundreds of them were pressed into action. Each sortie of a fighter aircraft may cost about 10,000 to 20,000 dollars (excluding the cost of weapon launched) and 2000 to 3000 sorties per day were flown. Which country other than USA and or a combine of some advanced countries can afford such deployment of frightfully expensive weapons or weapon systems ? Therefore it is not likely that in the near future any country would challenge the USA to war or would dare accept a challenge of war from the USA. Saddam till sixteenth January 91 was hoping that the USA would not enter the battle field if his (USA's) allies were not threatened.

One more factor has to be kept in view when discussing future wars. Earlier it was possible to compensate for the lack of technology by numbers, using defensive techniques, innovative tactics and high morale etc., but now onwards it would not be so. Earlier, if the interceptor aircraft was lower in performance than that of attacking aircraft, a larger number of interceptors could have defended the target from an attack by superior fighters, may be accepting a higher attrition than generally accepted. But now it can not be so. Once the radars and communication of one side have been jammed then the number of 'blind and deaf' SAMS or interceptor aircrafts would not matter.

Countries other than the USA and its allies may indulge in a war, with conventional but not so hi-tech weapons and therefore it would not be a hi-tech war of the calibre of Kuwait War. However, if any of such a country has got any nuclear capability, then the situation is wrought with grave dangers as either deterrent may not be effective, or a desperate leader may be mad enough to use that nuclear weapon. This is what demands that nuclear non proliferation treaty (NPT) be made fair to everyone and then accepted universally. In any case, gradual destruction of nuclear weapons must be carried out. This is not so much to avoid a war, as to avoid catastrophic results of a nuclear war. However, if two nations, not so developed, engage in a nuclear war, they both would suffer hell and the whole world would force them to stop the war inconclusively. Thus, such a mad war, though hypothetical, could not be called a hi-tech war.

War in Distant Future

Future is always difficult to discuss as it remains so unpredictable, and the more distant the future, more difficult it is to predict. Even then efforts are always made to look into future. Theoretically, one may argue that it can not be accepted that no country will ever catch up with the USA in the development of weapons. For example, it should be possible to develop certain electronic counter-counter measures that would render the jammers less effective, and it should be possible to develop radars that can see a stealth aircraft and so on. Then the balance would tilt in the favour of defending country so that the attacker would have to pay a heavy price before he can establish an air superiority which may remain either partial or may play 'cat and mouse' game. It may be possible to develop laser jammers and IR jammers which can reduce the accuracy of laser and IR guided weapons. Work is going on, in these directions, in the USA and some of the advanced countries like the UK, Germany, Russia and France etc. But is it going on effectively in developing countries, remains doubtful, if for no other reason than lack of suitable priority. For a developing country to develop weapons of the caliber of F-117 A, Tomahawk, SLAM etc. is a pipe dream; but they can certainly attempt to develop suitable electronic and laser counter measures and counter-counter measures to reduce the effectiveness of such high-tech weapons. Remotely piloted vehicles are also weapon systems of future and can be developed by comparatively less rich countries. But one thing is definite, a technologically poorer nation has no chance to win a war in future except when the war is between tweedledom and tweedledee.

No discussions of future wars can be complete without considering satellites and radiation weapons. Although these two weapons do not appear to be any candidates in the foreseeable future, it is simply because there is no one to challenge the Super Power. However it would still be interesting to dwell on the subject, briefly. It has been described that satellites are powerful means of communication, surveillance and navigation - all the three services vitally important in any war big or small. Indeed early warning radars have been positioned in satellites to warn the USA whenever an ICBM is launched against it. Such satellites are called early warning satellites. Therefore, in a war, satellites, specially those which are used for military purposes, become critical targets and that is the beginning of 'Space War' or 'Star War'. Satellites can be attacked electronically by jamming their electromagnetic spectrum, they can also be physically attacked by appropriate missiles or by satellites themselves

Some satellites have small jets with their motors, that are used for fine tuning their orbit from ground control. Enemy can use these jets, by controlling them, (provided they can find out the method of controlling,) to dislodge them from their correct orbit. In such a method satellites can be made to enter back into earth's atmosphere and get burnt, or to go far beyond in to space. Cosmonauts can walk in space whilst they remain tied to the mother satellite by means of a long umbilical cord. Now it is possible to produce a space vehicle for one person (you may like to call it 'space scooter') with the help of which a cosmonaut can travel longer distances away from the mother satellite. Such units are called 'man manoeuvering unit (MMU). With the help of an MMU, a commando cosmonaut can go close enough to a victim satellite and fire a gun (hand carried) at it to destroy it and then return to the mother satellite.

Jamming of geo-synchronous satellites, electronically speaking , is not very difficult except that one would need a very large and manoeurveable antenna to focus the jamming beam precisely on the victim satellite. A large antenna, say 10 to 15 m dish can be seen by a surveillance satellite and then can be attacked by missiles, say a cheaper version of Tomahawk. Jamming of low orbit satellites would also not be easy because it has to be acquired visually or electronically when it rises on the horizon and then tracked by a ground system with a tracking antenna. Frequency agility and other spread-spectrum techniques right now are so superior that the e.m. transmission can not even be received by unwanted persons therefore having acquired a low orbit satellite, the task then for knowing the frequency of its electromagnetic waves would still pose certain difficult problems to solve which would need rather large amount of jamming powers. In any case jamming and ECCM are always evolving like a 'sword and shield syndrome' and right now it is the shield which is somewhat stronger than the sword, as far as spread spectrum techniques are concerned.

Death-ray has been the dream of dictators, generals, scientists and science fiction writers. Tremendous time , energy and money has been and to some extent still being spent on the research and development of death-ray, the scientific name being radiation weapon. There are two types of radiation weapons viz. 'laser beam' and 'charged particle beam'.

Laser beam as everybody knows is a monochromatic beam of coherent (all the rays in the beam propagate in same phase) rays of light. Laser gets its power, because it can be focused into an extremely narrow beam whose power density, not the total power, can be made so high as to melt even iron and diamond and thus drill a super fine hole through them. But there is still need to increase the power density so as to make its destruction power devastating and , at the same time, keeping the weight and size of the laser producing system (laser gun) small enough to be usable.

Charged particle beam is a beam of high velocity electrons (negatively charged fundamental particles) or protons (positively charged fundamental particles) and a proton is about 2000 times heavier than electron. Proton guns, being more powerful, are preferred over electron guns.

A powerful proton beam can disintegrate metal objects. Although the mass of each proton is unimaginably small (micro-micro-micro-micro grams), but the numbers used in a single shot of proton gun is also unimaginably large and so is the speed, thereby their energy content is also very high. This energy of protons is measured in electron volts (the electric charge of a proton is same as that of an electron except that it is positive) and the energy of each proton when fired from a gun is of the order of billion electron volts. Proton gun is also bulky and cumbersome and therefore work is going on to make it practically usable.

Laser and proton guns would be very suitable for, inter alia, intercepting missiles specially ICBMs, launched by an enemy. Laser beam has an advantage that it can be reflected by special mirrors, placed in suitable orbits around the earth. Therefore when an ICBM/IRBM after its launch is not visible from ground due to earth's curvature, a laser beam can be shot at that ICBM/ IRBM by using a suitable mirror. Particle beam can be used only when 'line of sight' to the target is available . On the other hand , a particle beam, when of right strength, would kill only human beings and not destroy the equipment or buildings etc. Power of a laser beam gets reduced by smoke and dust particles present in its path. It may appear that death-ray would be a supreme weapon that could not be countered, but history tells us that a weapon and a counter weapon or defence have a dialectical relationship. Human brain would invent a suitable counter to radiation weapons, provided he survives.

If hi-tech wars of future are to be discussed, weapons described above ought to be used in that war. What are the countries that are likely to have those weapons. The USA definitely and may be Russia are likely to have those weapons developed in future. It can be safely stated that Russia is not in a position to enter into war with the USA in this century. In the past various States in Europe have given rise to many wars, big and small. Now with a move towards European Union, chances of a big war to begin in Europe have become remote. If a war breaks out among some non-European nations, it would not be hi-tech war of the calibre of the Kuwait war. China is the only theoretical candidate which can afford to go to war with the USA. Fortunately, China is also gradually moving towards 'free market' and there are no serious causes which can make a war between them imminent. Therefore we felt justified in giving the title to this book ' The Hi-tech War of Twentieth Century'.

Of course, as it stands today, there may not be any need to use radiation weapons or for that matter any bombs or rockets to dominate another country. This is so not only because there is only one super power, but also because the magical 'electronic media' and 'free market' together have made 'trade' and

---

[In every TV there is an electron gun, obviously it is not so powerful as to destroy the glass screen of the TV, yet at such a low power, these electron beams in a TV generate X-Rays (weak) on their impact with the glass screen. One should not watch TV from a very close distance.]

culture' as highly desirable 'weapons'. Surprisingly, not only attacker finds these weapons as desirable, but also the 'victims'. Trade is not a zero sum act like a war where one gains only at the cost of other. Trade, no doubt, benefits both the parties, but weaker party always gains less. One must remember that trade is a weapon which favours strong parties, companies and nations. Economically and technologically weak nations would always be dependent upon the strong nations who would like to exploit the weaker nations through their 'high voltage' (hi-tech?) sales 'techniques'.

It may not be much of an exaggeration, if it is said that the future war' is one in which the place of lethal weapons like Tomahawks and F-117As, the invisible bombers etc. has been taken by Coca Cola, jeans, Mickey Mouse etc. These are the symbols of marketable products of Western consumer culture and media which promote mindless consumption of products of Western technology. It should not be surprising that the same electromagnetic waves and electronics (the high priest of hi-tech) which decided the victor in Kuwait war are working as the media in this war also, and probably with similar effects. Jeans and Michal Jackson, Coke and American Soap Operas, Micky Mouse and rat race, foreign aid and AIDS, valueless free market and free sex, all these are new colonisers who colonise the minds of their victims. While teaching them lessons on 'freedom' and 'democracy', if the attackers succeed, the result will be that everybody will be equal at the alter of consumption - there would be no differentiation in the temple of God Consumption, no brother, sister, wife, son, father, mother - everybody a consumer, free to consume, as per his 'power' (not as per his need). Since we can not eliminate a bloody war, we can 'sublimate' it into a consumer festival till we consume the mother earth.

On this earth, the 'stronger' has almost always exploited the 'weaker', be it women, men, companies or nations. Undue exploitation may generate wealth for some but does not generate happiness, on the contrary it generates dissatisfaction and enmity and leads to confrontation. We have had, fortunately, a Mahatma Gandhi amongst us, but we are not yet civilised enough to understand and follow his practically demonstrated message of Truth, Love and Non- violence. Do we have any hope in the near future ?

Peace is Continuation of War With Other Means

Failure in the statesmanship results into a war as war is continuation of policies with other means. Use of strength in enforcing one's own policies or desires on others is as old as the history of mankind. Alexander The Great, Napoleon, Hitler had no excuse for a war except their ambitions!! It reminds us of Shakespeare who said : "and the big wars, that makes ambition virtue". To avoid a war and settle the disputes with negotiations is rather a newly born concept. Nearly equal strength of both the sides acts as a major deterrent to wars. Possibility of a tough resistance also discourages war. Until the world becomes a family, we have to remember the Greek saying, "Let him, who desires peace, prepare for war". But when there is not likely to be any worthwhile resistance and a clear and a quick victory could be achieved through modern and sophisticated weapons, the more powerful side may be tempted to wage a (high-tech) war rather than getting engaged into civilised negotiations towards achieving an objective which may be difficult to justify to the other party.

Specially since WW II, peace has been continuation of war by other means! And now, trade would be the weapon of war to maintain the so called peace and superiority of advanced nations. There is no solution now except to become technologically and economically strong through dedication, dilligence and wisdom without becoming a mindless consumer.

Power Corrupts & Absolute Power Corrupts Absloutely

Remember that power corrupts and absolute power corrupts absolutely. Now it looks that there is only one super power in the world and there is nobody to 'strike' a balance. There is some hope as both the big powers have agreed to reduce their weapons. All the developed nations are cutting down their defence budgets, and there is open working democracy prevailing in the country of the super power. John F Kennedy, the President of USA, had proclaimed in 1961, "Let the word go forth from this time and place, to friend and foe alike that the torch has been passed on to a new generation of Americans - born in this century tempered by war, disciplined by a hard and bitter peace, proud of our ancient heritage". Even then, is it not imperative that the UNO should be made more powerful and provided with a better democratic structure. Kuwait War has proved that the concept of Multi-National Forces can be effective. Either UNO should be provided with their own armed forces on a permanent basis or on ad-hoc basis which could be decided after various deliberations. But its principles should be laid down now, during peace time. Whether to provide armed forces to UNO or not may be debatable or may be agreed with a proviso of UN becoming more democratic but there cannot be any doubt that UNO should be made more democratic and more powerful.

At present the international laws on war, do not accept the oft quoted adage, 'all is fair in love and war'. War crimes include crime against humanity, genocide, and violations of the laws and customs regulating the conduct of war. Planning and waging of aggressive wars is also a crime. Use of force is limited by the UN charter to self defence as set out in Article 51, which provides that a state has " ..the inherent right of self defence..", a right that continues until the UN security council takes necessary action to secure peace. The law forbids e.g., using the flag of truce or Red Cross as a cover for hostilities, torture of captives. Espionage is permissible, use of weapons to cause unnecessary suffering and deaths is prohibited, biological/chemical warfare is forbidden, etc. etc.

War laws are commendable, specially because they intend to civilise the conduct of war. However nothing much can be done against the erring party till the war is over. Firstly, because the UN has no real force, like a nation can use against its criminals. Secondly, the whole issue is extremely complex, including the legal definitions. Thirdly, the UN security council does not really represent the world body in the true democratic sense. At present, a just action against a war criminal state is more of an exception than a rule.

UNO, despite its shortcomings, has been successful to some extent in bringing peace to this conflict ridden earth. We must remove those short-coming. Therefore, if we wish to control wars and make them less savage, the first action is to make the functioning of the UNO and the UN security council more democratic. Secondly, we must then give some real powers to the UN which they can use to punish the guilty. Thirdly, to oversee the functioning of the UN itself, even when it has became practically democratic, a UNO judiciary system should be evolved, but again with necessary power to enforce its decisions. At present, the UN court at the Hague can only pass strictures against a defaulting nation. Such a judicial system is fit only for a highly civilised society and not yet for the present lot of people inhabiting this planet earth. Many cases can be quoted. CIA mined the waters of Nicaragua which complained to the international court of justice. In order not to face the judgement, the US did a step better than to disregard the court's strictures, the US withdrew from the jurisdiction of the court. Only when some teeth are given to the UN court, will the UN be able to act as per its noble charter which develops three inter- related approaches: first, peaceful settlement of disputes; second, collective security to confront aggressors with too much to fight against; and third, disarmament.

In order that we are able to change the UN, apart from diplomatic efforts, efforts to inculcate human values in the world's citizens are also necessary. This means along with 'free market' values, we must propagate human values[1] such as

---

1.   Indian rishis had practiced and propagated such human values in distant past, as depicted by the couplet; " Ahimsa satyasteya brahmcharyaparigraha yamah". This means that non-violence, truth, not desiring what belongs to others, restraint of sexual urges and control of excessive indulgence in senses are five paramount disciplines for a human being, aspiring for mental and spiritual peace.

cooperation, consumption only to satisfy needs, simple living and high thinking, no greed, respect for other human beings, specially women should not be treated as a commodity for consumption; truth, ahimsa and love as values in all citizens. Free market without such human values converts a man into a consumer animal. And even if we are able to ensure no-war among nations, what would it achieve, if all of us men, women and children are at 'war' with each other in order to 'consume' the maximum.

Before we, in India, put up such a proposal to other countries, is it not a pre-requisite that we should ourselves function as a true democratic nation consisting of citizens with human values as mentioned above ! Is not GANDHI still very relevant ? Fear, though very natural, when beyond our capacity to cope with, dehumanises us. Ahimsa is victory over fear, over violence, a march towards sanity and wisdom, towards higher civilisation that homo sapiens are capable of. We have, from time immemorial desired, invoked and worked for brotherhood ('Vasudhaiva Kutumbakam[1]') and peace - "Vanaspatayah Shantihi, Prithivi Shantihi, Dyau Shantihi, Antariksha Shantihi[2], "

> Politics is continuation of war with other means.
> Peace is continuation of war with other means.
> Ahimsa is continuation of civilisation with peaceful means.
>
> — V.M.T.

---

1. Whole world is a family.
2. Let peace prevail in nature, on the earth, in the sky, in the cosmos.

Appendix

Electromagnetic Spectrum

Radio or TV programmes are received through special type of waves called radiowaves. Radio waves, heat radiations from fire, light-rays from a lamp or sun, X-rays, gamma rays and cosmic rays all belong to the energy spectrum known as electromagnetic (e.m.) waves. True to the nomenclature, these waves have electric as well as magnetic properties. When radiowaves emanated by a transmitting station reach the domestic radio/TV antenna, due to their electromagnetic properties they generate electric currents in the receiving antenna which finally reach the radio/TV receiver through the connecting wire/cable and we are able to listen/see the transmitted programmes.

E.m. rays basically contain electromagnetic energy in the form of rays of particles or waves. Waves could be characterised by three parameters viz. amplitude, phase and frequency (see illustration). Waves are also generated in water of sea or rivers and they could help us in understanding the properties of waves. The height of a wave (from its position in rest) is termed as amplitude. Every wave has crests and troughs. The distance between any two successive crests or troughs is called wavelength. These waves move and the distance they travel in one second is called their velocity. The number of full waves or the number of crests or troughs present in that distance (which is travelled in one second) is called the frequency of that wave, i.e number of times per second the formation of crest-trough-crest takes place. If a small piece of wood is placed in the water where

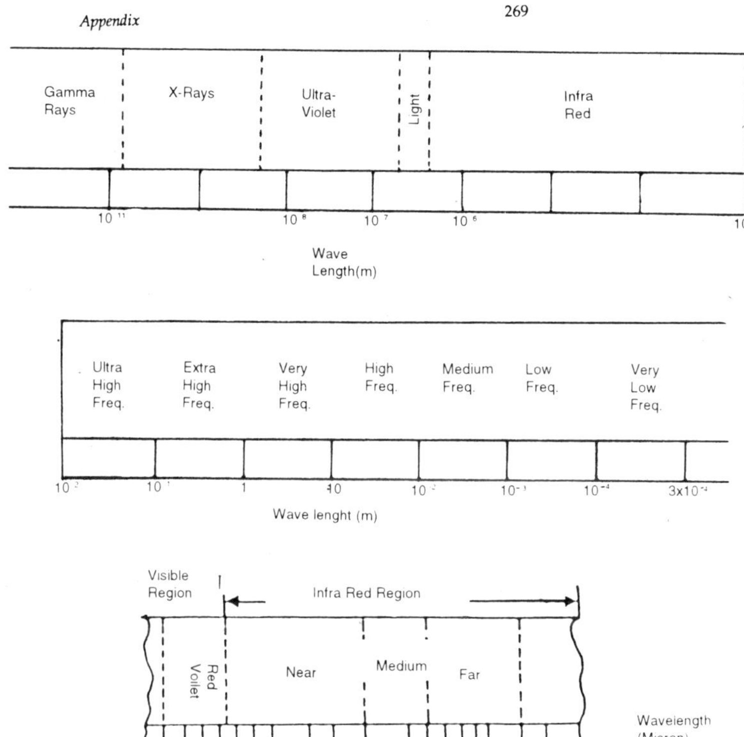

Figure 39
Electromagnetic Wave Spectrum

waves are being formed, the number of times the wood piece goes up and down per second would be same as the frequency of the waves. During the process of formation of two successive crests the wave height first reduces upto trough and then increases till the next crest is formed. Hence, one could say that there is a range of height levels between the two successive crests of a wave. These positions are identified by their respective phase or the phase angle. For the convenience of measurement, the distance between two consecutive crests is divided into 360° and these angles determine or denote the phase of a particular point. Thus there would be 180° phase angles between a crest and its nearest trough because a trough is exactly at the centre between two crests. As crest and trough are exactly opposite to each other, the term 180° out of phase for two waves means that the two are totally out of phase with each other i.e. one wave is at its crest, the other is at its trough. If these two waves have same frequency and same amplitude, then if they are superimposed on one another, the two waves would cancel each other and there would be no wave left at all.

When e.m. waves travel, no vibrating matter (like water for waves in a pond) is essential. In sound waves, the air particles vibrate to and fro (along the direction of travel) and not up and down (perpendicular to the direction of travel) as in waves on water, whereas in e.m. waves only the electric and magnetic intensity goes on increasing and decreasing. The e.m. waves travel with a speed of 3×10^8 meter per second in space (in vacuum), which is a universal constant. The velocity of these waves reduces while travelling through transparent or semi-transparent materials like water or glass.

The properties of e.m. waves completely change with the change in their frequency. The e.m. waves having frequency little lower than that of light waves have heat properties rather than those of light. The e.m. waves having frequencies higher than those of light have properties of ultra-voilet and X-rays etc. In fact the different colours in light waves correspond to different frequencies. Thus 'frequency' (or wavelength) of an e.m. wave is its most important characteristics which determines its type of behaviour.

Laser

The word laser is the acronym for 'Light Amplification by Stimulated Emission of Radiation'. What happens when two rays of light with equal amplitudes interact with each other ? Whether the intensity or amplitude of the resulting ray would double ? This would depend upon the other two properties viz. frequency and phase, if the frequency of both the rays are same, it would depend only upon their phases. It may also be possible that the relative phase difference between the two light rays· is such that at a particular instance one of them has a crest and the other has a trough, and as their frequency and amplitudes are same, the two rays of light would interact to cancel each other and produce darkness. This is an interesting case when two light rays meet and create darkness.

When a lamp or bulb glows, it emanates innumerous rays with no definite phase relationships. Many rays with opposite phases would neutralise each other whereas those with the same phase would add and increase their intensity. The rays produced by an ordinary lamp/bulb have a mixture of different colours of light i.e. mixture of different frequencies. The relative phase difference of waves of different frequencies do not remain same, they go on varying. In such a case our eyes perceive the mixture of two waves (colours) as a third colour. When two or more e.m. waves of the same frequency have zero phase difference, they add up to produce more intense light and such rays are called Coherent (phase correlated) rays. A laser beam is made of phase correlated rays of the same frequency (monochromatic). Laser beam could be produced by solid (synthetic ruby), liquid or gas (carbon-dioxide) matter. The molecules of the matter are stimulated by super high frequency waves and thereafter that special material like ruby produces monochromatic light rays. Due to the phase correlation in the constituent rays, the intensity of laser beam remains at theoretical maximum. And while travelling, a laser beam diverges very slightly ($0.15°$), unlike ordinary light rays. They, rather move like soldiers marching in steps, only in one direction (without dispersion). Due to this reason, unlike the normal light, the laser beam does practically suffer no attenuation (reduction in intensity), while travelling in vacuum.

Radio Receiver

It is absolutely essential to have a prior knowledge about the radio frequencies of the enemy's communications when electronic jammers have to be used to disrupt their tele-communication circuits. As you would have seen that on the same radio set you can listen to numerous transmissions by tuning the receiver, as different radio stations use different frequencies in such a way that they do not interfere with each other. However, when two programmes, by mistake, use the same frequency for transmission, they inadvertently cause disturbance to each other. Work on development of radio receivers, which could quickly find out the radio frequencies of the enemy communication systems, started during the WW I itself because all that is required is a radio receiver which can be tuned to different transmissions and whose frequencies could be measured.

Direction Finder

When enemy's communication system is to be disrupted or completely broken down with minimum jamming power, it is also essential to know the precise direction of the enemy trans-receivers. Once the direction is known, all the radiated jamming power could be concentrated only in the particular direction. Such uni-directional jamming would simultaneously reduce the possibility of disrupting other communication systems operating at the same frequency but located in other directions. Therefore, we need to have a direction finder and a directional antenna for achieving effective jamming. It is interesting to know that Marconi developed the first long range signalling system in 1901 which was deployed over the Pacific Ocean, and within four years, in 1905, he got a patent for developing a direction finding antenna. Again in 1912 he took a patent for another direction finding system. Such a system could find out the exact direction of the radio emitters. The technology of direction finding has advanced a great deal since then. Now the directions can be determined in fraction of a second and with accuracies of one degree or better.

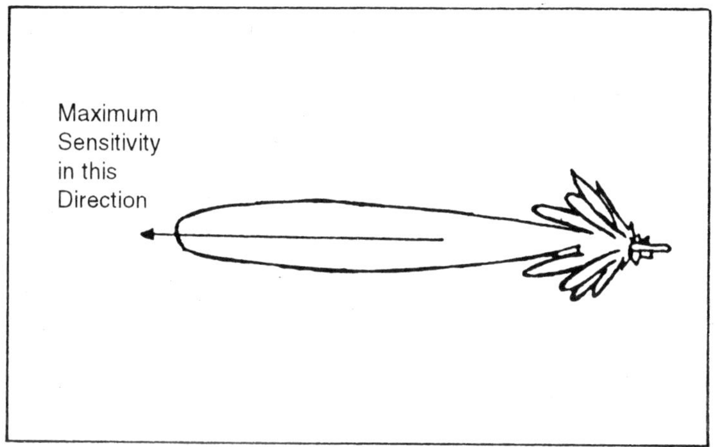

Figure 40
Radiation Pattern of a Direction Antenna

Radar

Radars have the capability of not only seeing the distant objects during day and night under all weather conditions, they can also measure the distance and direction of a target. Using e.m. waves a radar can estimate the distance and direction of any object which could sufficiently scatter or reflect the e.m. waves. From this point of view, aircrafts and ships, both are amenable to be detected, located and tracked by a radar. With the help of an antenna, a radar transmits e.m. waves in the form of series of pulses in a narrow beam in any specific direction. The narrow beam is formed due to the large directional antenna which is an integral part of a radar system. The large antenna may have the capability to rotate in azimuth as well as in elevation to search the targets in all directions. A surveillance radar antenna goes round on its axis and searches targets in all the azimuthal directions. When an e.m. pulse from the radar encounters any metallic object, like an aeroplane, the same pulse is scattered back towards the radar also. The radar receiver receives the pulse and computes the total time taken by the pulse to reach the object and return to the radar. The distance of the object is then evaluated by multiplying the time with the velocity of e.m. waves and then dividing the same by 2. As continuously changing distance is to be measured (for any moving object), a series of pulses are sent. Radar system has a transmitter and an antenna to transmit such a series of pulses. The same antenna is used for reception purposes also. The location of the target i.e. the direction and the distance thus measured by the system is displayed on the radar scope (a display screen like a TV has).

It is worth mentioning that immediately after the invention of wireless (radio) communication by Marconi in 1896, Christian Hulsmeyer, an engineer from Dusseldorf, in 1904, successfully demonstrated a ship locating system and also applied for its patent. However, at that time none of the shipping companies showed any interest in such an instrument. Sometimes the capabilities and farsightedness of engineers and scientists are far ahead of their times.

During the year 1922, Marconi had mentioned about radar like system. Two American physicists, Gregory Breit and Merle Tuve, in 1924, used a radar like equipment to measure the height of ionospheric layer (70 miles) in the atmosphere. They also successfully demonstrated a similar radar system, on a smaller scale, during 1933. Since then the research in the field of radars got accelerated. During 1934, Dr. Rudolf Kühnold of Germany developed a radar system, with the help of which he measured out the distance and direction of a ship sailing 7 mile away. While tracking the ship, he could also evaluate the distance and direction of an aircraft which by chance had come that way. In 1935, the first practical radar system was designed and developed by a British scientist, Watson Watt. The continuous research and development in the field resulted into modern radars which are not only capable of locating distant ships/aircrafts but can also provide suitable directions to artillery guns for correct aiming at the designated targets. Thus a radar serves the purpose of an electronic eye' for a ship or a weapon launcher. The way our eyes are the most important sensory organ of our body, a radar is the most valuable part of an air defence system. Hence, it is obvious that, tactically, the radars are the most important targets in electronic warfare.

Tracking Radar

If we watch the radar scope for sometime, we will find the dot representing the aircraft (blip) moving slowly. The highly directional radar antenna rotates horizontally on its axis to search objects in all the azimuthal directions. The antenna rotation speed generally varies between 6 to 60 rotations per minute. In each rotation the location of the blip of the aircraft changes due to the change in the distance of the aircraft from the radar. In the illustration given here, the aircraft has moved 6 km in 20 seconds. Such an estimation that the aircraft has travelled about 6 km in 20 seconds was done after watching the movement of blips on the radar scope. If there are large number of aircrafts in the area, it would be very difficult, if not impossible, for the radar scope observer to keep a track of all the targets. Therefore, electronic systems have now been developed which could do the task of identifying and tracking a target (aircraft or ship) automatically on command.

The speed of the moving target is also estimated with the help of Doppler's effect. Precise and accurate computations are performed by electronic systems to correlate the direction and speed of each target and track them continuously. All the time they have the exact knowledge of each target's position and they make use of such information. The acquisition radars use the information in respect of threatening targets themselves; and also provide it to the radars which control the artillery fire or

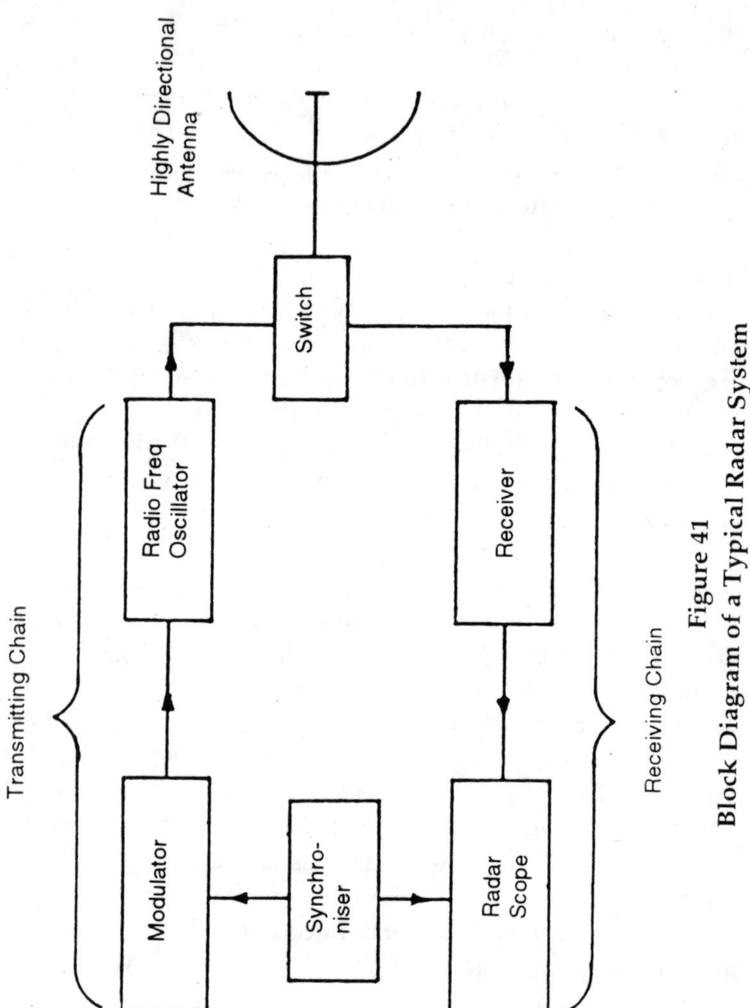

Figure 41
Block Diagram of a Typical Radar System

provide necessary information to other guided missile systems etc.

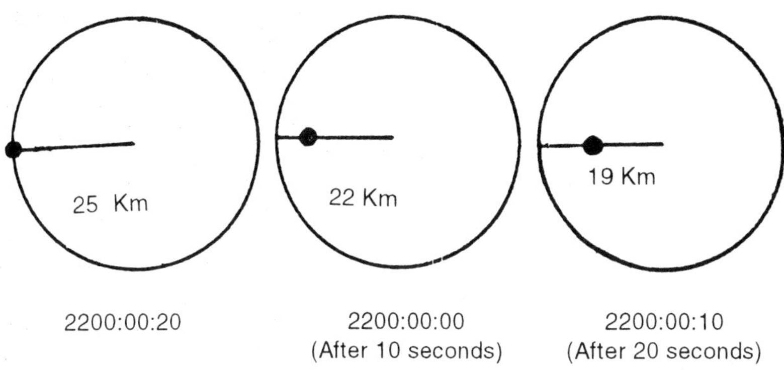

Figure 42

Radar Guided Missile

For missile guidance applications, the radar should have information about the position of the missile vis-a-vis the designated target, on a continuous basis. After obtaining these two informations continuously, the guidance radar computes and decides the exact direction in which the missile should move and accordingly it (radar) continuously sends suitable instructions to the missile's direction control sub-system. This guidance method is known as command guidance.

Fully Active Missile

In another type of guidance, the complete system of target tracking radar, the associated antenna and processing subsystem etc. is an integral part of the missile itself. This system provides the direction and distance information of the target to the missile to enable it to automatically guide itself on its own. Such a system makes the missile bulky and heavy but on the other hand increases its accuracy as well as its immunity to electronic counter measures. It is worth noting that the fully active guidance missile, after launch, could independently guide itself on to the pre-determined target accurately. After the second World War, Neil Cameron, the former Chief of Air Staff, Royal Air Force (Britain) had predicted that with technological advancements, the complexities and guidance capabilities of the weapon platforms would get incorporated in the weapons.

Semi-Active Missile

In the semi-active guidance, the missile does not contain complete radar system as its integral part. The missile contains only a receiver and its antenna whereas the radar transmitter (the bulky component) and its antenna are located in a ground or airborne platform. The radar beam radiated by the ground based/airborne system is reflected by the target and this reflected radio energy is received by a radar receiver located in the missile which then evaluates the direction and distance of the target vis-a-vis the missile and thus provides basis for self guidance. This system also makes the missile heavier but less than that of the fully active missile. Semi-active principle is generally used in the air-to- air (AA) and air-to-surface (AS) radar guided missiles since the fighter or attack aircrafts, in any case, have airborne radars for other purposes. These radars therefore, can be used to transmit an e.m. beam which after reflection from the target can be received by the AA or AS missiles.

Laser Guidance

Laser guided missiles function mostly in a semi-active mode and make use of a laser beam instead of a radar beam. The aiming accuracy of the laser guided missile is far better as the laser beam is extremely narrow in comparison with the radar beam. In some cases laser guidance is performed in the command guidance mode also.

Infra-Red Guided Weapons

The three main guidance systems discussed above are of active mode type as they all have to radiate electro-magnetic waves. The infra-red (IR) guidance is of passive type as no radiations are emitted by the guidance system. Since every object, depending upon its temperature, radiates thermal rays, such rays can be received and used for target aiming purposes. Such systems have IR sensing receivers to provide the direction of the target and help the weapon in homing on to the target. Characteristically these weapons strike at the hottest spot of the target. Such weapons have electronic stealth capability also as they are not radiating any e.m. waves, except their own thermal radiation.

Infra Red Imaging Guided Weapons

A common IR receiver can only sense the direction of the thermal source and provide information for homing on to it. Now such sensitive IR sensors have been developed which can differentiate objects with as low a temperature difference as 0.1 degree celsius. With the help of such sensors a thermal image of any object (target) could be taken and shown on a TV scope. Any point on the picture could be tracked. The missiles, based on such guidance technique, could home-on to strike the target at any desired point with precision - during day and night-under any weather condition, in smoke or fog etc. Maverick and SLAM missiles employ this technique of guidance. Such weapons also have electronic stealth characteristics.

Digital Scene Matching

Any picture taken by a video camera could be digitised and stored in an electronic memory. Countries like USA and France use different satellites, fitted with special cameras, to take digitised pictures of various important regions of the world and store them in electronic memories i.e. chips and not magnetic tapes. Spot Corporation, a private company of France, sold the digitised maps of certain parts of Iraq to US Air Force during Kuwait war.

Such digitised imageries were used by Tomahawk missiles for its terminal guidance. The stored memory of the guidance system of the missile contains the digitised map of the target area. As soon as the missile reaches the target area, its camera starts taking pictures and these digitised pictures are compared with the pre-stored digitised map of the area, thereby determining its position accurately and continuously. The missile strikes the point as soon as the target is reached with the help of the two digitized scenes. In our day to day life whenever we see any scene again, we recognise it as so and so only by comparing the present scene with the old one already stored in our memory. The same concept is used in these missiles. This technique also provides stealth capability and its counter measure is extremely difficult, e.g. a large area would need to be covered with smoke.

Gyroscopic Guidance

In this method, small but very heavy flywheel is rotated on its axis with a very high speed. The two ends of the axis on which the wheel rotates, are mounted on a metallic ring (say no.1). This metallic ring is also free to rotate on an axis perpendicular to the previous one. The two ends of the axis of the metallic ring no.1, are on another metallic ring (no.2) in such a way that the axes of both the rings are perpendicular to each other. The two ends of the axis of the second metallic ring are again on another ring. The axis of the third ring is not only perpendicular to the axis of the second ring, but also to the axis of the first ring. In other words the axes of all the three rings are perpendicular to each other, as shown in the illustration.

The heavy wheel in the gyroscope is rotated with a high speed and hence it is called a flywheel. Suppose, in the beginning of the experiment, the flywheel's axis is kept in the North-South direction, the axis of the second ring is kept in East-West direction and the axis of the third ring is kept vertical, and the flywheel is rotated with a very fast speed. Now if we rotate the base of the third ring in any direction, contrary to common

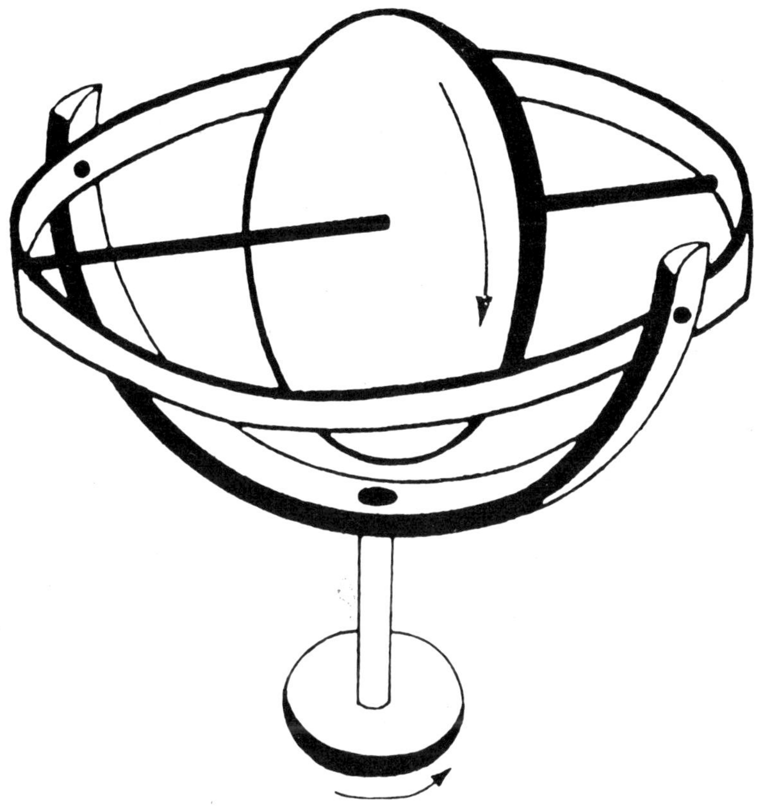

Figure 43
Controlled Gyroscope

sense, the flywheel's axis will continue to remain in the North-South direction. If we mount such gyroscope in a missile or an aircraft then the axis of the gyroscope flywheel would remain in the specifically pre-set direction, inspite of the change in the direction of the missile or aircraft platform on which the gyroscope is mounted. Such a characteristic of flywheel is called gyroscopic inertia. When all the three axes are independent of the housing body, the instrument is termed as 'free gyroscope', but they do not find much practical application in movement control or for guidance purposes. Instead a 'controlled gyroscope' is used for various functions. In such a system, by some suitable method, the flywheel axis is kept in a specific direction with respect to the earth, say North-South, (and not with respect to space, as in the case with free gyro), so that a reference direction is available for use, all the time, despite any movement of the housing body i.e. an aircraft or a missile. Such an instrument in an aircraft is called a gysocompass. Apart from direction, gyros are used to indicate 'rate of turn' and rate of roll (specially for aircrafts). Gyros are also used for auto-pilots or for missile guidance (e.g. Scud missiles).

Scud Guidance

Errors are caused due to the friction in the support of axes therefore the gyroscopic guidance may not provide high order of accuracy. However, they are still better than the unguided (dumb) rockets and could be used for large size targets like factories etc. A modified version of such a system (as used by 'V2' rockets) was used in Scud, Al Abbas and Al Hussein missiles. The 'V2' rockets fired by Hitler on London during World War II were provided with gyroscopic guidance system.

Terrain Contour Matching

Gyroscopic guidance is preferred even today for long range missiles in their initial phase of travel, specially when it is over sea. There are many reasons for this. Firstly, it is difficult to get any navigational point over the sea surface ; secondly, the application of active radar type of technique would negate its stealth property; and thirdly, the small error generated during the first phase of travel could be rectified during the next phase of travel. A cruise missile (like Tomahawk) launched from a ship reaches land after travelling some distance over sea.

For navigation during the second phase the three dimensional maps of the travel route are used. The depth of all the points falling on the travel route are noted with respect to the height at which the missile or aircraft is expected to fly. If these depths are plotted on a plain paper, we get a terrain line diagram of the travel path which is called 'contour cross- section' map. Now, if an aircraft or a missile during its flight keeps on measuring its height (by using a radar altimeter) above the surface of the earth, it also can get a contour cross-section of the path traversed. If the terrain of the flight path does not mostly contain sea or desert, the contour cross-section for each flight path would be unique. With the help of three dimensional pictures taken through special satellites, the contour cross-section of any flight path could be obtained and stored in the memory of the missile's computer. When the missile traverses the same path, using the height finder radar, it could draw the contour cross-section of its flight path. If on comparison the two cross-sections are found matching, the correctness of the flight path is ensured. The terrain contour matching system has a limitation which depends upon the number of contour plots it can store in its memory. It is not capable of making any alterations in the flight path of the missile if it goes very much off the desired routes stored in the memory. Contour matching may take long time unless one is flying close to the desired path. For such a contingency gyroscopes are used to bring the missile or aircraft approximately on the desired path. The errors in the gyroscopic guidance are precisely rectified by the contour matching technique.

Spread Spectrum Technique

Satellite communication like any radio communication is vulnerable to interception and jamming by the enemy. Therefore, it is provided with ECCM capabilities to counter the threats from various ECM attacks. The messages are coded and in addition, 'spread spectrum technique' is used for such communication to provide resistance to both interception and jamming attempted by enemy. To understand the spread spectrum technique, one has to first understand the simple modulation principle. Sound is spread over a limited frequency band. The sound of music falls within a band of frequencies from about 20 Hertz (Hz) to 20 Kilo Hertz (KHz). Our voice on telephone line is generally contained in a band of frequencies from 300Hz to 3000 Hz. The sound energy at the audio frequencies, as it is, can not be transmitted over long distances for the purpose of communication, as they suffer severe path loss and high powers cannot be generated and the resulting noise pollution would be intolerable. For transmission over long distances, the audio information (after converting the sound energy to electrical energy) is 'superimposed' on e.m. frequencies of higher bands (radio frequencies) to enable them to travel long distances. This special 'superimposition' is correctly termed as modulation : The radio frequencies are called carrier frequencies and the information being superimposed is called modulating signal. When the amplitude of the carrier wave is changed according to the modulating signal the process is termed as amplitude modulation. Likewise, in case of frequency and phase modulation, frequency and phase of the carrier wave respectively are changed according to the modulating signal. These basic modulation techniques, however, do not provide any security to the communication. Such signals could be intercepted/jammed with the help of suitably designed systems which are in the realm of simple technology.

To introduce a strong element of security in the radio communication, some additional processing is done on the modulated signal and one such process is called 'spread spectrum modulation' (SSM). The word 'modulation' in SSM is used only for technical reasons as it is quite different from the objectives for which the modulation techniques are generally used.

In SSM, (spread spectrum modulation) the information (message) is spread over a large bandwidth, very much larger than conventionally required for transmission of information. The spreading of the bandwidth is done through a secret code which is known to the authorised users only. To extract the

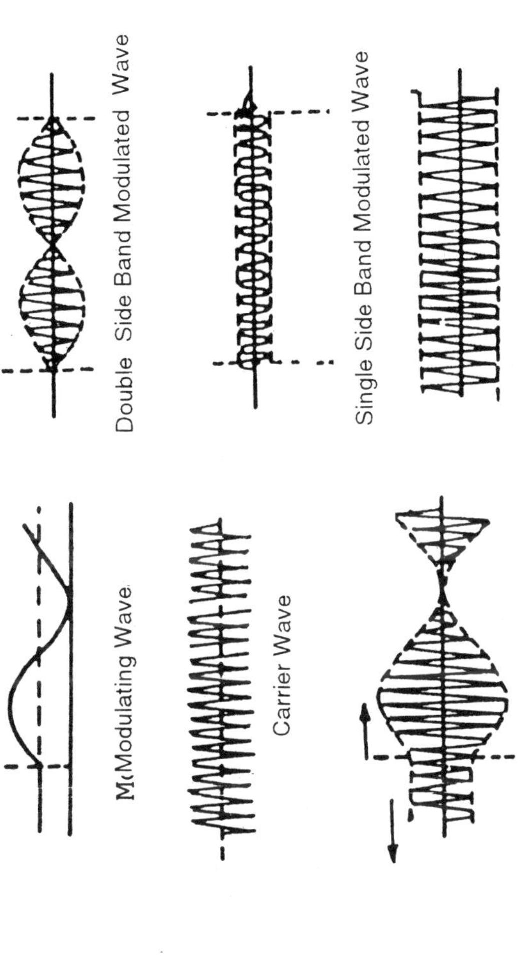

Figure 44

Amplitude and Frequency Modulation

original information, the SSM signal could be demodulated only if the secret code is known. This code could be in the form of 'pseudo random noise' or 'pseudo random bit sequence'. The spreading of the bandwidth brings down the signal power density (power per Hertz) and it is desirable to bring it below the level of the natural noise present in the atmosphere. If the power density of the signal before spreading was 40 times higher than the noise (which is appropriate for proper reception), it may be brought down 25 times lower than the noise level i.e. a reduction of the density of signal power by 1000 times. Since the signal is now fully immersed in the noise, a conventional radio receiver, however sophisticated, cannot receive the signal and extract the information due to predominance of noise and absence of the correct code for demodulation.

To achieve still higher order of security the modulation codes could also be changed from time-to-time. In another method of SSM, the frequency of transmission in the given band could be changed at a very fast speed (called frequency hopping) in a pseudo random manner, the generation code (hopping sequence) of which is also kept secret. As the signal is being transmitted on different frequencies at different times, the unauthorised receiver would not know how to tune his receiver exactly at the changing transmitter frequency at different times.

The SSM technique also provides very high protection against radio jamming by the enemy. For effective jamming, the enemy has to increase the jamming noise power as per the ratio of the spreading unless he knows the secret code. Such ECCM techniques are used in satellite communication systems to achieve message security as well as jam resistance.

Some Military Terms

Strategy

1. **Clausewitz** : "The use of combat or the threat of combat for the purpose of the war in which it takes place".

2. **Jomini** : "All strategy is controlled by invariable scientific principles which prescribe offensive action to mass forces against weaker enemy forces at some decisive point to lead to victory".

3. **Heinrich Dietrich von Bülow** : "All military movements out of the enemy's canon range or range of vision".

4. **Moltke (the elder):** "A system of ad-hoc expedients, it is the application of knowledge to practical life, the development of an original idea in accordance with continuously changing circumstances".

5. **Hans Delbrück:** (a) Strategy of annihilation (Niederwerfungs- strategie) :" A war conducted with the sole aim of destruction of the enemy. (one pole war)". (b) Strategy of exhaustion (Ermattungs strategie) :" It is essentially an effective means of attaining the political ends and is therefore limited in its range of actions. It has two poles - battle and manoeuvre."

6. **General Lüdendorff (German General during later WW-I):** War is the highest expression of the racial will of life. Strategy becomes a form of social mobilisation which adjusts goals to means. Strategy's aim is to mobilise the nation for an unlimited war and is achieved through ideology rather than on rationality."

7. **Hitler:** "Strategy is the creation of fortuitous political circumstances that would allow the military to wage war." It became the main approach for shaping a world where basic inter- relations were based on struggle and conflict. Strategy, during Hitler's time was reduced to exploitaion of technology by ideology to win the war.

8. **General André Beaüfre:** War -"The dialectic of two opposing wills, using force to resolve their dispute". Strategy : "is the art of that dialectic". The object of strategy is to achieve and maintain one's freedom of action and to try and limit that of the enemy.

Blitzkrieg

It is a form of strategy in which speed, mobility and optimum use of weapons are used to deal a decisive blow on a weak point of enemy, may be by using outflanking movement.

Tactics

1.**Clausewitz:** "Tactics constitute the theory of the use of armed forces in a battle"

2.**Heinrich Dietrich von Bülow:** "All military movements within the enemy's canon range"

War

General Lüdendorff : "All theories of Clausewitz have to be thrown overboard. War and politics not only serve the survival the people, but war is the highest expression of the racial will of life."

Clausewitz: Continuance of policy by other means.

A rational limited instrument of national policy.

An act of violence intended to compel our opponent to fulfil our will.

Tolstoy: War is a calamity and a social disaster.

Modern thought: Armed conflict between political units, on a large scale (involving more than 50,000 combatants.)

Principles of War

Principles of war are those adherence to which have led to success in wars in the past. The generally agreed principles of war are:- The objective, the offensive, unity of command, concentration of forces, economy of force, manoeuvre (mobility) surprise, security from attack, simplicity and administration.

Introduction of Some Military Thinkers

Beaufre André, General: Author of three famous works : (i) An Introduction to Strategy {Paris, 1963) (ii) Strategy of Action (Paris, 1966) and (iii) Problems of Modern Strategy (edited by Alastair Buchan, I.I.S.S., London, 1980).

Brodie, Bernard: An American modern military thinker, who is a respected nuclear strategist. His book, " The Absolute Weapon" (1946) is also rated as a brilliant original work.

Clausewitz, Carl Von: A General in the Prussian army; One of the most widely read German military thinker, specially known for his epitome "On War" (1830s), Born: 1780 at Burg. He died during the great Cholera epidemic of 1831.

David Mac Isaac: After retiring as a Lt. Col. from the US Air Force, he worked as Senior Research Fellow at the Centre For Aerospace Doctrine, Research and Education, Air University, USA. He is the author of "Strategic Bombing in WW II".

Delbrück, Hans: Born 1848 in Bergen. Soldier, scholar, Professor and Parliamentarian. Author of "History of the Art of War", Editor: Preussische Jahrbusher.

Douhet, Giulio : Born 30 May 1869, Caserta, Italy. Died 15 Feb 1930. Army General, considered as father of strategic air power. From 1912 to 1915 he served as commander of the Aeronautical Battalian, Italy's first aviation unit. At every opportunity, he recommended his strategic theories. His severe criticism of the conduct of the war resulted in his court martial and imprisonment. But investigation of the Italian defeat at

Caporetto in 1917 justified his criticism; his conviction was reversed, and he was appointed head of the Aviation Service. His most noted book is 'The command of the Air', 1942.

Fuller J.F.C.: Born 1878, Died 1966. British Army Officer (Colonel), military theoretician and war historian, known as one of the fathers of modern armoured warfare. He wrote many books: Tanks in the Great War (1920); The Reformation of War (1923); On Future Warfare (1928); Memoirs of an unconventional Soldier (1936); Machine Warfare (1942); The Second World War 1939-1945 (1948); A Military History of the Western World (1954-56).

Heinrich Dietrich von Bülow : A military thinker, respected for his interpretation of Napoleanic warfare.

Liddle Hart, Sir Basil (Henry) : Born 1895, Died 1970; Invalided in 1924. He retired as a Captain from the British Army in 1927. He became an early advocate of air power and mechanised tank warfare. He has written many books : The Remaking of Modern Armies (1927); The Strategy of Indirect Approach (1929); A History of World War 1914-18 (1948); The Other Side of the Hill (1948 and 1951); The Tanks (1959); Deterrant or Defence (1960); A History of World War II (1970).

Moltke (the elder) : (1801 -) He began service as a young Lieutenant in Prussian Army and rose to become a General in 1866. He continued his services as late as 1870-71, during the Franco- Prussian war.

Wasner, Edward : Aeronautical engineer, Professor at the M.I.T., Vice-chairman of the Civil Aeronautical Board, Assistant Secretary of the Navy for aeronautics, editor of 'Aviation'. His famous essay "Douhet, Mitchell, Seversky : Theories of Air Warfare" had appeared in 'Makers of modern strategy' (Princeton 1943 edition) and has been reprinted countless times in various magazines.

Index

Index